Have We Lost Our Minds?

"Stan Wallace has written an important—indeed, critical—book that clears away serious misunderstandings as well as misrepresentations regarding spiritual formation and neuroscience. Wallace uses his impressive philosophical skills and insights to cut through the fog and expose neurotheology's faulty conclusions. However, he also offers a constructive, biblical alternative to point us in the proper direction."

—**Paul Copan**, chair of philosophy and
ethics, Palm Beach Atlantic University

"I heartily recommend Stan Wallace's timely book—a fraternal but forceful critique of 'neurotheology,' the attempt by some evangelical Christians to redefine the human soul, mind, and spirit entirely as generated and animated by the brain. Wallace clearly explains why this implicitly materialistic perspective conflicts with Scripture, Christian doctrine, sound philosophy, proper science, and healthy spirituality. Instead, he encourages appropriation of sound neuroscience within the historic Christian perspective that affirms the basic distinctness and holistic integration of both body and soul."

—**John W. Cooper**, professor emeritus of philosophical
theology, Calvin Theological Seminary

"For many years, I felt settled in my understanding of the body-soul relationship. But now, to avoid substantial error, we must integrate new findings from neuroscience with our concept of the human person. Using Scripture, theology, philosophy, and working examples, Stan Wallace's book thoughtfully probes the body-soul-brain dynamic. From reading his thoughtful commentary, I have a deeper, richer understanding of what it means to 'belong, body and soul, in life and in death, to my faithful Savior, Jesus Christ.' So can you."

—**Shirley J. Roels**, executive director, International
Network for Christian Higher Education

"'Good philosophy must exist,' C. S. Lewis wrote, 'if for no other reason, because bad philosophy needs to be answered.' Because so much bad philosophy, both formal and popular, exists around the questions of what it means to be human, it must be answered. That's exactly what Stan Wallace gives us in *Have We Lost Our Minds*, along with a clear, articulate description of the beautiful, robust reality of the *imago Dei*, which is to say, who we truly are."

—**John Stonestreet**, president, Colson Center

"*Have We Lost Our Minds?* is a game changer! Sometimes we walk down an unsettled road for a considerable distance before a loving corrective voice sets us on the proper path. Stan Wallace does just that. Wallace brings a kind, gracious, scholarly perspective to move us in the direction of genuine human flourishing by confronting the errors of well-intentioned authors promoting neurotheology. I highly recommend this book."

—**Jimmy Dodd**, founder and president, PastorServe

Have We Lost Our Minds?

*Neuroscience, Neurotheology, the Soul,
and Human Flourishing*

STAN W. WALLACE

Foreword by J. P. Moreland

WIPF & STOCK · Eugene, Oregon

HAVE WE LOST OUR MINDS?
Neuroscience, Neurotheology, the Soul, and Human Flourishing

Copyright © 2024 Stan W. Wallace. All rights reserved. Except for brief quotations in critical publications or reviews, no part of this book may be reproduced in any manner without prior written permission from the publisher. Write: Permissions, Wipf and Stock Publishers, 199 W. 8th Ave., Suite 3, Eugene, OR 97401.

Wipf & Stock
An Imprint of Wipf and Stock Publishers
199 W. 8th Ave., Suite 3
Eugene, OR 97401

www.wipfandstock.com

PAPERBACK ISBN: 978-1-6667-8913-3
HARDCOVER ISBN: 978-1-6667-8914-0
EBOOK ISBN: 978-1-6667-8915-7

VERSION NUMBER 06/04/24

PERMISSIONS

Unless otherwise noted, biblical citations are from the New American Standard Bible®, Copyright © 1960, 1971, 1977, 1995, 2020 by The Lockman Foundation. All rights reserved. The "NASB," "NAS," "New American Standard Bible," and "New American Standard," are trademarks registered in the United States Patent and Trademark Office by The Lockman Foundation. Use of these trademarks requires the permission of The Lockman Foundation. For permission to use the NASB, please visit The Lockman Foundation website: www.lockman.org.

Scripture quotations marked (NIV) are taken from the Holy Bible, New International Version®, NIV®. Copyright © 1973, 1978, 1984, 2011 by Biblica, Inc.™ Used by permission of Zondervan. All rights reserved worldwide. www.zondervan.com. The "NIV" and "New International Version" are trademarks registered in the United States Patent and Trademark Office by Biblica, Inc.™

Scripture quotations marked (NLT) are taken from the Holy Bible, New Living Translation, copyright ©1996, 2004, 2015 by Tyndale House Foundation. Used by permission of Tyndale House Publishers, Carol Stream, Illinois 60188. All rights reserved.

Scripture quotations marked (GNT) are from the Good News Translation in Today's English Version—Second Edition Copyright © 1992 by American Bible Society. Used by permission.

Scripture quotations marked (CEV) are from the Contemporary English Version Copyright © 1991, 1992, 1995 by American Bible Society. Used by permission.

Scripture quotations marked "Phillips" are taken from The New Testament in Modern English, copyright 1958, 1959, 1960 J. B. Phillips and 1947, 1952, 1955, 1957 The Macmillan Company, New York. Used by permission. All rights reserved.

To Roger Hershey, who first modeled for me how to be a lifelong learner and follow truth wherever it leads, even if doing so is unpopular.

Contents

Foreword by J. P. Moreland | xiii
Acknowledgments | xvii

Introduction: Bodies, Souls, and Human Flourishing | 1
    The Historic Christian Understanding of What We Are | 6
    An Alternative Understanding of What We Are | 7
    Understanding and Evaluating Neurotheology | 8
    Mapping the Journey to Understanding | 11

1 **Neuroscience, Neurotheology, and the Soul** | 14
    Important Discoveries of Neuroscience | 15
    From Neuroscience to Neurotheology | 16
        Neurotheology's Interpretation: Our Thoughts
            Are Really Our Neurons | 17
        Neurotheology's Conclusion: Therefore, I
            Am Essentially a Brain | 18
        Neurotheology's Application: Neuroscience
            Is the Key to My Flourishing | 20
    Neurotheologians Are Physicalists | 20
        Are Neurotheologians Simply Using
            Complementary Explanations? | 22
        Are Neurotheologians Just Imprecise? | 24
    What Else Do We Know about What We Are? | 25

CONTENTS

2 **The Bible and the Soul** | 27

  Neurotheology's Biblical Anthropology | 29
  A More Adequate Biblical Anthropology | 31
    Our Body Is Crucially Important | 31
    We Also Have a Soul | 31
    The Body and Soul Are Deeply United | 34
    Yet Ultimately We Are a Soul That *Has* a Body | 35
  Christians Throughout the Ages Agree | 39
  Where Did Neurotheology Go Wrong? | 40

3 **Neurotheology's Wrong Assumption About our Mental Life** | 41

  Neurotheology's Wrong Assumption: What Identity Is | 43
  Mental Events Are Not Identical to Brain Events | 47
    Our First-Person Perspective | 49
    Our Free Will | 51
    Our Rationality | 54
  Are Neurotheologians a Nonreductive Type of Physicalist? | 55
  Inadequacies of Nonreductive Physicalism | 58
    Freedom Is Still Excluded | 58
    Rationality Is Still Excluded | 60

4 **Neurotheology's Wrong Conclusion about What We Are** | 63

  The Neurotheologian's Conclusion | 64
  Two Reasons Why This Conclusion Can't Be Right | 66
    Brains Cannot Explain Our Unity at Each Moment | 66
    Brains Cannot Explain Our Unity Through Time | 72
  Neurotheology's Wrong Application: How We Flourish | 77

5 **The True Nature of the Soul** | 80

  The Soul Is an Individuated Human Nature | 82
    What a Nature Is | 83
    What a Human Nature Is | 84
    What an Individuated Human Nature Is | 92

An Individuated Human Nature Is a Spiritual "Substance" | 93
  Substances Are Owners and Unifiers of Properties | 93
  Substances Are Enduring Continuants | 94
  Substances Change in Law-Like Ways | 94
  Substances Are Particular Things | 95

6 **The Unity of the Soul and the Body** | 96

Body and Soul: A Match Made in Heaven | 97
Unifying the Discoveries of Theology,
  Philosophy, and Neuroscience | 104
Naming This View: "Holistic Dualism" | 107

7 **Three Common Defenses of Neurotheology** | 112

Defense 1: Let Science Be Our Guide | 113
  Defining Scientism | 114
  Three Reasons to Reject Scientism | 116
Defense 2: Dallas Willard Was a Neurotheologian | 118
  Wilder's Interpretation of Willard | 118
  Why Wilder's Interpretation of Willard
    Is Deeply Misguided | 119
Defense 3: Neurotheology Is Helping Many People | 122
  Neurotheologians Are Helpful in Spite of
    Their Neurotheology | 122
  Why Neurotheology Is Ultimately Harmful | 129

8 **Three Common Objections to Holistic Dualism** | 132

Objection 1: Keep It Simple | 133
Objection 2: How Can The Body and Soul Interact? | 134
  Five Replies to This Objection | 135
  Three Reasons Why This Objection Persists | 140
  A Case of the False Dichotomy Fallacy | 143
Objection 3: What about Animal Souls? | 144

9 Soul, Body, and Loving God | 147

The Platonic/Cartesian Dualist Reduction to "Pure" Spirituality | 149
The Physicalist Reduction to Shaping Our Brains | 149
The Middle Way: Reducing to Neither Body or Soul | 150
    Spiritual Formation and Our Capacities | 152
    Spiritual Formation and Our Faculties | 156
    Spiritual Formation and Our Teleology | 158

10 Soul, Body, and Loving Others | 160

Loving Others as Christ's Ambassadors | 161
    Evangelism and Missions: Proclaiming the Good News | 161
    Biomedical Ethics: Fostering True Human Flourishing | 164
    Social Ethics: Pursuing Justice for All | 167

Loving Others Through Our Professions | 169
    Overall Approaches to Work | 169
    Loving Others Through Education | 170
    Loving Others Through Medicine | 171
    Loving Others Through Business | 172
    Loving Others Through Architecture | 172
    Loving Others Through Law and Politics | 173
    Loving Others Through Science | 173
    Loving Others Through Computer Science | 174
    Loving Others Through Vocational Ministry | 175

Conclusion | 178

Glossary of Technical Terms | 181
Suggestions for Further Reading | 192
Bibliography | 195
Selected Subject Index | 207

Foreword

IN SUMMER 1974, I attended a week-long campus ministry conference in Seoul, South Korea. My roommate was a Filipino layman sent by his church to attend the conference. After a restless night's sleep following our arrival, I awoke on the first morning of the conference. On my way to the shower, I noticed that my roommate was reading. I glanced curiously to see what the book was. To my horror, he was reading Rudolf Bultmann's *Kerygma and Myth*.

Bultmann was a German scholar known for his "de-mythologizing" approach to Scripture. He believed that although we could know that a man named Jesus existed, that was about all we could know. For him, the Gospels were largely myth and fabricated legend. Accordingly, Bultmann's book (which I had read myself) could be categorized as spiritual poison. My roommate's pastor, it turned out, was a secular liberal trained in a radical leftist European seminary, and he was teaching his congregation to follow a deconstructed New Testament in which all the supernatural elements were regarded as prescientific nonsense that should no longer be believed.

I knew I had to warn my brother about the spiritual impact of what he was reading, even though in doing so I would be casting aspersions on his pastor and other congregants who had bought into this non-Christian worldview. I was uncomfortable about doing so, but truth matters, and he was absorbing grotesque falsehoods that would lead him away from Christ. Therefore, I took the time to explain Bultmann's worldview and errors to him, urging him to see the flaws in these ideas and recommending works written by solid, Bible-believing thinkers.

In this situation, I was confronting ideas that were clearly contrary to biblical teaching. In other instances, the task of distinguishing truth from error is more delicate. Sometimes, in the interest of truth and biblical fidelity, we must confront the ideas of brothers and sisters in the Lord, especially when those ideas are fundamentally contrary to a Christian worldview. Over the years, I have felt compelled to do this many times. As uncomfortable as these corrective occasions have been, I had a moral and spiritual duty to correct what seemed to me to be important errors that at the core of historic Christianity.

For example, one year I attended the national meeting of the Evangelical Theological Society, and the plenary address was delivered by someone who defended open theism. Briefly, open theism contends that God is not omniscient and does not know the future actions of human persons endowed with free will. I believe this position is contrary to the biblical understanding of God. But I was even more aghast when the speaker appealed to the authority of the widely esteemed Christian writer Dallas Willard to bolster his case, claiming to the two thousand people in attendance that Willard was an open theist. I was deeply offended by the speaker's egregious misrepresentation of Willard's beliefs. After the session, I talked with this brother and graciously but firmly admonished him to correct the record and never again make such a claim about Willard. Taking this step was uncomfortable, but it had to be done.

Today, we find ourselves in a similar situation. As you will discover in the pages that follow, false and harmful (albeit well-intentioned) ideas about our fundamental nature as human persons are being adopted by pastors, Christian counselors, and laypersons in our churches. These ideas, purportedly based on recent advances in neuroscience, describe humans as fundamentally physical beings without an enduring soul—a view commonly referred to as physicalism. Dr. Stan Wallace has done a masterful, gracious, yet firm job of carefully and accurately exposing these harmful ideas as presented by two influential brothers with large Christian followings: James Wilder and Curt Thompson. Stan provides a penetrating critique of their ideas. But he doesn't stop there. He also presents a compelling biblical alternative, supported by some of the most important thinkers in the history of Christianity, that explains all the relevant facts of neuroscience, aligns with our spiritual experience, and has far-reaching practical applications.

Neither Stan nor I claim that Wilder and Thompson's writings are of no value. Among other positive things, they remind us that training and developing our body is an important component of growth in Christlikeness. They also provide useful justification for medication as a tool in dealing with various kinds of mental illness.

However, with these exceptions, everything of value in their books has to do with their discussion of distinctively spiritual, not neurological, aspects of growth. The information Wilder and Thompson have collected from neuroscience plays virtually no role in helping the reader in his or her spiritual pilgrimage. To discover this for yourself, read Wilder or Thompson, and whenever you encounter a section that identifies and describes the areas of the brain that are activated by various spiritual disciplines and practices, ask yourself this question: Exactly how is this technical scientific material helping me understand sanctification and grow in Christ? Other than the defense of appropriate medications and highlighting the importance of the body (which orthodox Christian thinkers have affirmed for centuries!), this material is not only unhelpful but even a distraction from the real issues involved in spiritual formation.

Moreover, although it pains me to say this, Wilder has committed the same error as the open theist I cited above: he has seriously misrepresented Dallas Willard. Wilder, in his book *Renovated: God, Dallas Willard, and the Church That Transforms*, depicts Dallas as a supporter of "neurotheology." Some personal background should help to explain why I am in a good position to understand what Dallas really believed. I completed my PhD studies under Dallas at the University of Southern California from 1982 to 1985. I took several courses with him, and he was my dissertation supervisor. From 1985 until his passing, Dallas and his wife, Jane, were very close friends; in fact, it sometimes felt as if they had adopted my wife and me as part of their family. We met frequently through those years, and I would call him every few months to catch up and talk philosophy. In short, I knew Dallas and his philosophical and theological thinking very, very well. I think it is safe to say that I am among the handful of people most intimately acquainted with him and his body of work.

Based on that knowledge, it is beyond my comprehension how anyone could portray Dallas as anything close to being a physicalist, a view he detested. Yet Wilder (perhaps unintentionally) egregiously misrepresents Dallas on this fundamental topic! Just as the brother in Christ who claimed Dallas was an open theist needed to be graciously admonished, in the interest of truth and respect for Dallas's insights, Stan does the same thing in this book. The church should be grateful for Stan's gracious but clear-thinking response to neurotheology and its misleading interpretation of human nature.

Publishing a critique of the popular ideas of fellow Christians is always a serious matter. I think all Christians would recognize that some disagreements are more important than others. Differences of opinion about God's existence, the deity of Christ, and the atoning sacrifice of Christ's death on the cross are serious matters; disagreements about the nature of church government, modes of baptism, and so on are important but less central to our faith. I would place Wilder and Thompson's ideas, because they concern our basic human nature and how we relate to God, closer to the former category (issues of central importance) than the latter category. That is why the body of Christ needs this book and why it should be widely read. Stan was uncomfortable writing it, because he does not wish to be divisive. But he knew he had to do it, to help believers remain faithful to very important ideas that are at the core of a Christian worldview and our understanding of humanity.

Some of the most fundamental debates taking place in universities and in popular culture today concern the nature of human beings. Now is not the time for Christians to be uninformed about this question, or to reinterpret Scripture to make us acceptable to our secular friends who embrace scientism. Stan's book is an articulate, rigorous, informative yet readable work for such a time as this. Theological drift is widespread in the Western church today, and Wilder and Thompson's neurotheology, though well-intentioned, is inadvertently contributing to that drift. Please understand that neither Stan nor I have any disrespect toward them as persons or brothers in the Lord. They love Jesus and sincerely desire to serve him. It is their ideas that we stand against.

I have carefully read and wholeheartedly endorse Stan's work. This is an important book. Because of the deep nature of some of the questions addressed, it is not always easy reading, although Stan has worked diligently to make complex ideas from philosophy accessible to general readers. Do not be dissuaded. Read it, give it to your pastoral staff and Christian counselors you know, and tell others about it every chance you get.

—J. P. Moreland, Distinguished Professor of Philosophy, Talbot School of Theology, Biola University and co-author (with Brandon Rickabaugh) of *The Substance of Consciousness: A Comprehensive Defense of Contemporary Substance Dualism* (Wiley-Blackwell, 2024)

Acknowledgments

My deepest thanks to J. P. Moreland, my mentor and friend for over three decades. I am grateful to him for first introducing me to the importance of thinking well about what we are, encouraging me to write this book, and giving me invaluable feedback on drafts along the way. J. P. has been a constant model to me over the years in his commitment to discovering what is true, living accordingly, and flourishing.

Bruce Barron has copyedited chapter drafts and applied his keen mind to the task of identifying inadequacies in what I was trying to say or how I was saying it. He frequently pushed me to find just the right word, phrase, or flow of argument.

Leah Frank, my editorial assistant, exemplifies this book's audience. She is a thoughtful believer, a seeker of truth, and a Christ-follower committed to living according to the Scriptures so as to love God and others well. My thanks for going above and beyond to prepare this manuscript for publication.

I thank Rob Coker for giving me input from the perspective of someone with a PhD in neuroscience as well as advanced training in philosophy of mind, and Patrick Norris for giving feedback as a pastor and a student of neurotheologians. My thanks as well to the many others who have sharpened my thinking over the years, including Scott Rae, Doug Geivett, Dave Twetten, Paul Copan, Paul Gould, Randy Newman, Cam Anderson, Mike Priest, Jerry Hertzler, Tim Howard, Mike Erre, Bob Sievert, Paul Young, Jim Dunphy, Chip Elmblad, and Chet Dickey.

I owe a debt of gratitude to the members of the Global Scholars board of directors, who encouraged me to pursue writing and provided the time and encouragement to see this project through to completion. It is a great honor to serve under their leadership.

Finally, my deepest thanks to my dear wife Lori, who provided extremely helpful editorial suggestions and was willing to let me steal away to my study for hours and even entire days at a time to write this book!

Soli Deo Gloria,
Stan W. Wallace
Olathe, Kansas
January 2024

Introduction
Bodies, Souls, and Human Flourishing

[Dallas Willard] had two main concerns. The first concern was that the spiritual formation movement be established on more intellectually rigorous philosophical and theological underpinnings.

—J. P. Moreland[1]

Extraordinary care must be taken to formulate correctly our understanding of humanity. What humans are understood to be will color our perception of what needed to be done for them, how it was done and their ultimate destiny.

—Millard Erickson[2]

> **CHAPTER SUMMARY**
>
> This introduction begins by explaining the historic understanding of persons as both a body and soul (or "mind," which is often used synonymously with "soul"). According to this view, we have both a material and an immaterial dimension, yet we are ultimately a soul that can live after our body dies. Yet some Christians believe that recent findings of neuroscience indicate we

---

1. "Dallas Willard Memorial Service, J. P. Moreland." Available at https://www.youtube.com/watch?v=AzSEeIUoksU&ab_channel=DallasWillardMinistries (from 4:00 to 4:36); accessed November 12, 2023.

2. Erickson, *Christian Theology*, 481.

> are only, or at least fundamentally, physical beings. As a result, their books advocate rethinking the nature of spiritual formation and human flourishing. Through the popularity of their writings, these ideas are increasingly prominent among Christians. Although I appreciate the pastoral intent of their books, these ideas must be evaluated in a way that takes seriously both the findings of neuroscience *and* what we know about our nature from Scripture, philosophy, and daily observation. I seek to carry out this evaluation in a simple but not simplistic way, making the conversation accessible to everyone interested in this topic. My ultimate goal is to provide a credible answer to the important questions "What am I?" and therefore "how do I grow in Christ and flourish?" and to show where the idea of us as fundamentally physical falls short and must be corrected.

IN 2021 I VISITED a very prominent Bible-believing church in my area. The pastor spoke on Romans 12:2: "Do not conform to this world, but be transformed by the renewing of your mind." I was glad he was teaching on this text, as the idea of loving God with the mind has fallen out of favor in many churches these days. But as he began, he immediately substituted "brain" for "mind" and spent the rest of his sermon talking about how our brains work in the process of spiritual renewal. To support this substitution, he repeatedly quoted Curt Thompson, a Christian psychiatrist and author of *Anatomy of the Soul: Surprising Connections Between Neuroscience and Spiritual Practices That Can Transform Your Life and Relationships.*

As I listened, I became increasingly uncomfortable, for in this passage Paul is speaking of renewing our *mind* (part of our immaterial dimension), rather than our *brain* (part of our material dimension). After the service, I shared my concern with the pastor. To help me understand the connection of neuroscience to spiritual formation, he suggested that I read a book by another influential Christian leader—Jim Wilder's

INTRODUCTION 3

Renovated: God, Dallas Willard and the Church That Transforms. Reading it raised more questions than it answered.

Just a week later, I was meeting with a longtime friend who ministers to business professionals. "Mark" said he was working on a book to help businesspeople serve Christ in their professional lives. He is usually very thoughtful in all he says and writes, so when he asked me to review a draft of the book, I was happy to do so. I was quite surprised to find that he too focused on understanding how the brain functions in order to best understand being faithful as a Christian in business. My concerns continued to grow.

Soon after this, I was asked to review the curriculum being used by a church planting ministry. It included a section on spiritual formation for church planters in training. Again, I discovered that the curriculum focused on our brain activity. It was based heavily on the writings of Jim Wilder.

Finally, I was invited to participate in a webinar hosted by a prominent ministry. The featured speaker was the previously mentioned Curt Thompson, who discussed how to help others grow in Christ. Once more the conversation revolved around understanding how our brains are the key to spiritual formation. The many people attending the webinar were eager to understand and apply these ideas in their ministry contexts. By this time I was very troubled, because it seemed that everywhere I turned, the idea being promoted was that we are ultimately a body, and most importantly a brain.

You may have been exposed to these ideas as well. Perhaps you heard them mentioned in a sermon at church, or discussed in a podcast. A friend may have brought these ideas up over a cup of coffee, or you may have run across them in a bookstore as you looked for something to read on spiritual growth.

Ideas like these matter. They can powerfully shape how we think, act, and live. This is why, in the passage cited above, the apostle Paul instructs us to "renew our minds." We must work constantly to understand what is really real, and then to live accordingly. Paul contrasts this honest approach to reality with the common human tendency to understand reality as the people around us—"this world"—say it is.

If we can resist the strong pull of the patterns of this world and truly live according to what is real, we will flourish. As Paul adds in Romans 12:2, if we renew our minds with true beliefs, we will "prove what the will of God is, that which is good and acceptable and perfect."

The question at the core of the interactions cited above is what we are at the most fundamental level. We say we are many things: a husband or wife, a father or mother, an employee, a homeowner. But none of these are essentially what we are. I was me before I married, had children, got a job or bought a home. Similarly, I could lose all these things tomorrow and I would still be me. Therefore, those things are not essential to what I am. Rather, I am something more fundamental that *has* these other things. But what is that something? What am *I*? Am I ultimately a body—a physical thing? Or am I ultimately a soul—an immaterial thing? Or am I ultimately some combination of the two?

A proper understanding of what we are is necessary for us to flourish, because what a thing is determines what it needs to thrive. Take the tree in my backyard. It is currently flourishing because it is planted in the right type of soil, with the right amount of moisture, nutrients, and sunlight. But if I dug a hole and planted my dog in the backyard with exactly the same conditions, my dog would not flourish! In fact, she would promptly die. This is because the nature of a dog calls for a very different environment than the nature of a tree.

In the same way, I will flourish only if I have a proper understanding of what I am. Important implications of the idea that we are ultimately material beings will be discussed in chapters 9 and 10. For now, I'll briefly touch on a few ramifications to illustrate the critical importance of properly understanding what we are.

As illustrated above, if we are ultimately a body, and most importantly a brain, spiritual formation is actually neural formation. Growth in Christ must be refocused on gaining a better understanding of how the brain is shaped. The training and success of pastors, spiritual mentors, and Christian counselors must also give greater attention to understanding neuroscience and the brain rather than understanding the soul and how it is formed.

If we are ultimately a body, evangelism and missions are not about the salvation of souls. They must be redefined in terms of enhancing other's physical lives. Concepts such as sin, Christ's incarnation, and his atonement must be understood as related to our bodies, not our souls. Even the idea of an immaterial realm existing at all—including such things as objective moral values—is less plausible if everything we encounter day in and day out, including other persons, are ultimately physical in nature.

Our professional lives will, in large measure, be determined by our understanding of what we are. If people are assumed to be ultimately physical, the emphasis will be on meeting physical needs alone. Take, for instance, the medical professions. If we are ultimately material beings, it follows that all ailments must ultimately be physical, and therefore the interventions prescribed should also be physical.

Our understanding of people as ultimately physical also has wide-ranging implications for our cultural values—how we believe society functions best. If we are ultimately material, the highest cultural value is the freedom to meet our physical needs, as we define them. In our consideration of biomedical issues such as abortion, if we are ultimately physical we must define life in terms of specific bodily functions such as brain activity or responsiveness to stimuli, with significant implications for when we understand life to begin and end. Finally, if we are ultimately physical beings, there can be no such thing as intrinsic value, fundamental equality, justice for all, or inalienable human rights, for there is literally nothing we all share in common that could ground these values. Taken to its extreme, this reductionistic view of what we are undergirded the Holocaust, for instance.

As these examples illustrate, for us to thrive—to live as we hope to live, and as God has called us to live—we must understand *what* we are. Ultimately, this understanding will help us grow in our walk with Christ (love of God) and love our neighbors as ourselves (love of others).

The first eight chapters of this book will explore what we are from both a biblical and philosophical perspective, leading to a more detailed discussion of these points of application in chapters 9 and 10. I realize that diving into theology and philosophy can be challenging. Yet, as is often true, error is simple but truth is more nuanced. Therefore, it is important to explore a bit of the nuances of our souls and bodies in order to have the robust understanding of human nature necessary to better understand our growth in Christ.[3] If you are more of a "show me the money" type of person, you may want to jump directly to chapters 9 and 10 first, to read more about the implications of a proper understanding of what we are. Then you can come back and work through chapters 1 to 8 in order to "backfill" these implications with more of the biblical and philosophical framework.

---

3. Deepening our theological and philosophical understanding of reality helps us discern truth from error in all other areas of our lives as well. Therefore, this book is a case study in how to "think Christianly" about all of life.

THE HISTORIC CHRISTIAN UNDERSTANDING OF WHAT WE ARE

Throughout the ages, most Christians have understood Scripture as teaching that we are a unity of two dimensions: a body (a material dimension) and a soul (an immaterial dimension). More specifically, Christians have viewed human beings, in this life, as embodied souls—immaterial beings that in some way are united with physical bodies. I will briefly summarize the biblical teaching here. Chapter 2 will then examine this in more detail.

The idea that we are both body and soul is grounded in biblical passages such as Jesus' warning, "And do not be afraid of those who kill the body but are unable to kill the soul" (Matt 10:28).[4] The debate over the relation between body and soul is quite complex; we will return to this topic in greater depth in chapter 6.

Furthermore, most believers have understood Scripture as affirming not only that we have two dimensions, but that *both* our material and immaterial dimensions are important aspects of what we are. The reality and value of the material realm, including our bodies, has been upheld against contrary views such as Gnosticism, the early distortion of the Christian faith that treated the body as irrelevant or even evil, because the Gnostics believed that all matter was evil and that only spiritual things are good.[5] From this, they inferred our bodies are of no value. Furthermore, they argued that Jesus could not have had a physical body at all, because a material body would forever taint him and therefore he would not be God.[6] The opening words of 1 John respond to this Gnostic idea: "What was from the beginning, what we have *heard*, what we have *seen* with our eyes, what we have looked at and *touched* with our hands, concerning the Word of Life."[7] John argues that contrary to the false teaching of the Gnostics, Jesus indeed had a material dimension, and that this was an important aspect of his existence in the world. Since Jesus was like us

---

4. For a detailed evaluation of this passage see Cooper, *Body, Soul and Life Everlasting*, 117–19.

5. Gnosticism as a fully developed system of thought did not appear until the second century. In 1 John and elsewhere biblical writers were responding to an early, incipient form of "proto-Gnosticism."

6. For a discussion of how Gnostic philosophy influenced the early church and gave rise to 1 John, see Stott, *The Letters of John*, 48–51.

7. First John 1:1 (emphasis added).

in every way (Heb 2:17), it follows that our bodies are also an important aspect of what we are and how we flourish.

Yet the traditional understanding of biblical teaching is that our immaterial dimension—our soul—is even more foundational to what we are, because this dimension continues to live after our bodies die. As Paul expresses it, "While we are at home in the body we are absent from the Lord, . . . but we . . . prefer rather to be absent from the body and to be at home with the Lord" (2 Cor 5:6, 8).[8] Therefore, since we can exist without the body but not without the soul, the soul is ultimately what we are. In the words of C. S. Lewis, "A soul is that which I can say I am."[9] It follows that caring for the soul is essential to our flourishing.

As illustrated earlier, a growing number of Christian pastors, counselors, psychiatrists, and authors are implicitly or explicitly challenging this traditional understanding of what we are. They are stating or implying an alternative understanding of our nature.

AN ALTERNATIVE UNDERSTANDING OF WHAT WE ARE[10]

Due to recent findings in neuroscience, a number of influential believers are suggesting that our material dimension—the body—is ultimately[11] what we are. I will draw on the two prominent books I mentioned above—*Anatomy of the Soul* and *Renovated*—to illustrate how this alternative is often implicitly assumed, and sometimes explicitly stated, in the context of discussing human flourishing.[12]

Thompson and Wilder argue that until recently, we had only a fragmentary understanding of what we are. But now, due to the advances of

---

8. See Cooper, *Body, Soul, and Life Everlasting*, 141–48. Chapter 2 will discuss the historic biblical understanding of the soul and body in more detail.

9. Lewis, *The Collected Letters of C. S. Lewis: Narnia, Cambridge, and Joy*, 10. For more developed arguments for the soul as the self, see Duvall, "From Soul to Self and Back Again," 6–15; Moreland, "Restoring the Substance to the Soul of Psychology," 29–43.

10. Chapters 3 and 4 will explore this alternative understand in much more detail.

11. I use "ultimately" or "fundamentally" here and throughout to allow for the possibility that they may also believe we have a non-physical dimension (including mental states such as thoughts and beliefs). This variant will be discussed in more detail in the following chapters. Yet even if so, as will be discussed in chapter 1, it is clear that they believe this immaterial dimension emerges out of the more fundamental thing that we are—a brain.

12. Wilder and Thompson have both written a number of other books, which echo similar themes.

neuroscience, we can much more fully answer this question. As Wilder summarizes, he seeks to "examine whether current brain science would change the understanding of human nature that has dominated Christian theology since the Middle Ages."[13] Thompson adds, "The fields of psychiatry, genetics, developmental and behavioral psychology, psychoanalysis, neurology and neuropsychology, developmental neurobiology, and functional neuroimaging ... add to our understanding of how we have come to be who we are."[14] Yet until recently, "knowledge from the many scientific fields has not been integrated into a single coherent body of knowledge."[15] Thompson rejoices that now scientists are beginning to offer us this integration, in such forms as Daniel Siegel's "interpersonal neurobiology."[16]

This approach to integrating the findings of neuroscience and the theology of spiritual formation has become known as "neurotheology." Wilder defines this pursuit as "the science of spiritual maturity."[17]

Given the growing prominence of neurotheology, we must be careful to follow the Lord's admonition in 1 Thessalonians 5:21: "Examine everything; hold firmly to that which is good." I serve with Global Scholars, a ministry that equips Christian professors to be the "aroma of Christ" (2 Cor 2:15) in higher education globally. In my role, I regularly encourage Christian professors to consider how they may serve Christ by engaging the ideas in their fields of study from a biblical worldview. In what follows, I will attempt to practice what I preach, examining the ideas of neurotheology from my areas of expertise so as to determine what is true and therefore what is right and good to believe.

UNDERSTANDING AND EVALUATING NEUROTHEOLOGY

Some may be tempted to write off neurotheology without much thought, since the Scriptures seem to teach clearly that we are souls that possess bodies (including brains). But Christians must also affirm science as a means of discovering what is real, for God surely reveals truth in his

---

13. Wilder, *Renovated*, 2–3. Wilder offers no support for his claim that the traditional understanding of human nature dates back only to the Middle Ages. I will argue in chapter 2 that the traditional view of humans as body *and* soul has dominated Christian understanding from the time of the biblical authors.
14. Thompson, *Anatomy of the Soul*, 5.
15. Thompson, *Anatomy of the Soul*, 6.
16. Thompson, *Anatomy of the Soul*, 6.
17. Wilder, *Renovated*, 127.

creation (his *general revelation*). "The heavens declare the glory of God; the skies proclaim the work of his hands. Day after day they pour forth speech; night after night they reveal knowledge" (Ps 19:1–2 NIV). God has given us science to help us understand things as they are through the study of his creation. Therefore, we must always be open to what new discoveries tell us. Perhaps Christians have misunderstood for centuries what we really are. If so, neurotheology may be giving us a clearer understanding of our nature and therefore how we flourish. Anything we can discover about this topic, from any field of study, should be of great interest to us.

Accordingly, I deeply appreciate Thompson and Wilder's desire to help believers integrate all we know about what we are, including what we can learn through neuroscience, to aid us in our spiritual formation. As Wilder asks, "Would knowing how the brain learns character revise how we teach ourselves to be Christian?"[18] Thompson concurs, stating that by understanding neuroscience we can better understand "why we do what we do over time."[19] Therefore, in his book he introduces "several neuroscientific concepts that have great significance to the community of faith."[20] Wilder states similarly, "Reconciling the church's practices of transformation to how the brain works will be our topic for this book."[21] This pastoral concern is evident throughout their books. As Thompson puts it in his epilogue:

> My work involves helping people pay attention to the elements of their [brains] . . . and then integrating these disparate parts so that we can live a life of mercy and justice in every realm and dimension of life together. I believe God's Kingdom advances when this integration occurs in the community as well as in the individual.[22]

These are most admirable goals! So, as Thompson and Wilder are dear brothers in Christ, discussing a very important topic with good intentions

---

18. Wilder, *Renovated*, 3
19. Thompson, *Anatomy of the Soul*, 5.
20. Thompson, *Anatomy of the Soul*, 6.
21. Wilder, *Renovated*, 7.

22. Thompson, *Anatomy of the Soul*, 257. I put "brains" in brackets because here he uses the term "minds," but it is clear that he is referring to the brain, He uses these terms synonymously throughout the book, as I will demonstrate in subsequent chapters. Furthermore, on page 9 he states that "the terms brain and mind . . . are . . . closely enough related to seem interchangeable."

and motives, we must give their perspective an unbiased evaluation by clarifying precisely what they are claiming and then fairly evaluating their ideas.

On the other hand, some may be tempted to embrace the ideas of neurotheologians without much thought, assuming that these ideas are based on science and therefore *must* be true. However, this reaction is as problematic as ignoring neuroscience and neurotheology altogether. Science is certainly one way to know truth. But it is not the only way.[23] We know that God also reveals truth through his Word (his *special revelation*). Biblical scholars and theologians therefore have much to offer on issues such as this which are addressed in Scripture. We dare not discount the knowledge gained from their studies. Furthermore, philosophy also discovers truth by studying God's general revelation. After centuries of exploration, philosophers have also gained much knowledge about what we are and how we flourish. Their knowledge must not be discarded either.

Therefore, since "all truth is God's truth,"[24] I wish to find the middle way between the two extremes of fully embracing or fully rejecting contemporary understandings of neuroscience. Our study must certainly consider what we know from neuroscience, but it must also include what we know from theology and philosophy. Only as we integrate *all* we know about what we are from these three domains of knowledge will we be able to develop a true and full understanding of what we are and how we flourish.

I am also seeking a middle way in a second sense. On one hand, this topic is of great importance, and so I hope my discussion is not so superficial as to be of no help in discerning the truth of the matter. I am haunted by words that were spoken at Dallas Willard's memorial service: "[Dallas] had two main concerns. The first concern was that the spiritual formation movement be established on more intellectually rigorous philosophical and theological underpinnings."[25] I want, insofar as I am able, to contribute to the depth of understanding Willard had and others continue

---

23. The limits of science will be discussed in more detail in chapter 7.

24. This phrase is a common summary of Augustine's argument, "Let every good and true Christian understand that wherever truth may be found, it belongs to his Master." *On Christian Doctrine*, II.18, available online at https://www.ccel.org/ccel/augustine/doctrine.xix_1.html, accessed July 21, 2023.

25. These comments were made at Willard's memorial service by his longtime friend and protégé Dr. J. P. Moreland. Available at https://www.youtube.com/watch?v=AzSEeIUoksU&ab_channel=DallasWillardMinistries (from 4:00 to 4:36), accessed November 12, 2023.

to develop, which is so necessary to develop an increasingly robust understanding of spiritual formation and, more broadly, human flourishing.

Yet on the other hand, the fields of theology and philosophy have developed very technical terminology over millennia of discussing this issue. And more recently, as neuroscience has become its own field of study, it too has developed a rich vocabulary. As much as possible, I will avoid these technical terms and nuances so as to make the discussion accessible to the non-specialist. When such terms are important, I'll offer a definition the first time they are used, and I will include these terms in the glossary. For those who want to go deeper, at various points I'll provide footnotes with additional terms, issues, nuances, and suggested books that may be helpful for further study, as well as a list of useful sources in the Appendix.

MAPPING THE JOURNEY TO UNDERSTANDING

Chapter 1 begins this exploration with a brief summary of recent discoveries in neuroscience that Thompson and Wilder use as their foundation. As my point is not to challenge the findings of neuroscience (in fact, I applaud these scientific advances), the majority of chapter 1 then outlines the contours of neurotheology, due to their interpretation of the scientific data of neuroscience. My focus is on the neurotheologians' understanding of us as ultimately physical beings, which becomes the foundation of everything else they promote. As is true of anything we build, neurotheology either stands or falls on the basis of this foundation.

Chapter 2 surveys what the Bible has to say about what we are. Indeed, both the Old and New Testaments offer a lot of information on this topic. The Bible portrays us as everlasting souls that are deeply united with our bodies. This results in a deep functional unity of the two. Though we will be separated from our bodies at death, our body and soul will be reunited at the final resurrection and we will then live as embodied persons forevermore. This understanding is then shown to be consistent with the interpretation of Scripture on this topic throughout the centuries.

Chapter 3 considers why the understanding of what we are offered by neurotheologians differs so drastically from the picture that emerges from Scripture. The neurotheologian's error is traced to their fundamental assumption that when a neural event and a mental event occur together, the two must be identical. This assumption is evaluated and shown to be unfounded, based on three counter-examples: our first-person subjectivity, our free will, and our reason. Finally, an alternative form of physicalism, which Thompson and Wilder may embrace, is identified. After evaluation this alternative form of physicalism it is found to be of no more help in defending the anthropology of neurotheologians.

Chapter 4 shows how the wrong assumption discussed in chapter 3 leads neurotheologians to the erroneous conclusion that we are fundamentally a physical thing—ultimately a brain. This understanding of our fundamental nature is shown to be inconsistent with two features of ourselves that we know to be true: our unity at a time and our unity through time. Finally, based on the neurotheologian's wrong conclusion of what we are, I show how this leads to their inaccurate application to questions of human flourishing.

Chapter 5 explores in more detail the true nature of the soul, drawing on what we can know from philosophy. An answer emerges that echoes the biblical text: we are an individuated human nature, or a "spiritual substance." Philosophical insights also help us make connections to what we learn from neuroscience and theology, as discussed in chapters 1 and 2.

Chapter 6 discusses how the soul relates to the body and outlines how this view is consistent with what we know from theology, philosophy, and neuroscience. The various terms used for this view are discussed, including my preferred title, "holistic dualism."[26]

Chapter 7 considers three defenses neurotheologians may offer in support of their view in response to my criticisms, as outlined in chapters 2 through 6: (1) science must be our guide, (2) Dallas Willard endorses neurotheology, and (3) neurotheology helps many people. After finding these three defenses inadequate, I conclude that although there is certainly some truth in what neurotheologians say, and although they have helped many people in spite of (*not* because of) their neurotheology, ultimately their understanding of what we are, and its implications, are quite harmful.

Chapter 8 evaluates three arguments neurotheologians may offer against holistic dualism: (1) neurotheology offers the simpler answer to the question of what we are, (2) souls and bodies are so different they cannot interact, and (3) holistic dualism seems pantheistic in promoting

---

26. Here and elsewhere when the term "dualism" is used, it refers to *anthropological dualism*: views that understand the human person as having a body and a soul. This is not to be confused with the many other ways the term is used in other contexts.

the idea that animals also have souls. These three arguments are evaluated and also found to be inadequate.

Chapter 9 moves to application, discussing how we can best love God in light of what we have learned. I consider how what we know about the soul and its relation to the body helps us develop a better understanding of the nature and process of spiritual formation in relation to our soul's capacities, faculties, and teleology (its natural end or goal).

Chapter 10 applies what we have learned about the soul and body to loving others in two contexts. First, I discuss how holistic dualism helps us love others as Christ's ambassadors as we seek to share the gospel and promote the common good. This is illustrated both in relation to biomedical ethics (using abortion as a case study) and social ethics (using the idea of "justice for all" as a case study). Second, I apply this understanding of what we are to how we can best love others through our professions, using eight occupations as examples: education, medicine, business, architecture, law and politics, science, computer science, and vocational ministry.

Finally, in a brief conclusion I suggest how we should think and speak of our human nature so as to value the findings of neuroscience but not reduce us to fundamentally a brain. I end with an invitation to use the approach offered in this book as a model for other areas in which we must integrate all we know to answer the important questions of our day.

If you are especially interested in the ideas of neurotheologians such as Thompson and Wilder, as well as critiques of what they are endorsing, chapters 1, 2, 3, 4, 7, and 8 will likely be of greatest interest to you. If you most want to understand what we are and how we flourish, you will find chapters 2, 5, 6, 9 and 10 more helpful.

May God give us, in the words of the illustrious German astronomer Johannes Kepler, the ability to "think God's thoughts after him" as we explore more fully what we are and how we can "put on the new self, which is being renewed in knowledge in the image of its Creator" (Col 3:10). This is one of the most important journeys we can take, and it will have far-reaching benefits!

1

Neuroscience, Neurotheology, and the Soul

We honor God for what he conceals; we honor kings for what they explain.
—Proverbs 25:2 (GNT)

The *left hemisphere* [of the brain] sets me apart as *"me."*
—Curt Thompson[1]

> **CHAPTER SUMMARY**
>
> This chapter begins by affirming some important discoveries of neuroscience about how the brain works. Two important discoveries are highlighted: the connections between the brain and thinking, feeling, and our other activities, and the fact that the brain can be shaped—its "neuroplasticity." The way in which neurotheologians interpret this data is then evaluated. They interpret the data of neuroscience to mean that the mind and brain are the same thing, and that therefore we

---

1. Thompson, *Anatomy of the Soul*, 244 (emphasis added).

> are essentially our brains. From this, they conclude that we must understand the brain better, via neuroscience, in order to understand spiritual formation and human flourishing. I explain that their anthropology is in effect a form of physicalism. The chapter concludes by illustrating the importance of also considering what we can know about ourselves from other points of view, specifically theology and philosophy.

GOD TAKES GREAT DELIGHT in the study of his creation. One way in which we study God's creation is through science. As science explores the physical aspects of the world God has created, one important part of this investigation is our own physiology. This includes the study of how our brains work. The field of neuroscience is committed to this task. With the help of increasingly sophisticated technology, neuroscientists are making great strides in developing a better understanding of the brain's intricacies.

IMPORTANT DISCOVERIES OF NEUROSCIENCE

Overall, neuroscientists are making two important discoveries. First, they are finding clear correlations between regions of the brain and our mental life (thinking, desiring, choosing, and so forth). Second, they are discovering that our brains can be reshaped in order to function better; this concept is known as "neuroplasticity."

For example, a popular book discussing various interpretations of neuroscience begins by listing the results of a number of important studies (it is not important to understand the technical neuroscientific terms used[2]—the point is to illustrate the cause-and-effect relations being discovered by neuroscientists):

---

2. As mentioned above, I'll define scientific terms that Wilder and Thompson use in their reasoning as they arise. For helpful definitions of many other neuroscientific terms, see the National Center for Biotechnology Information's *National Library of Medicine,* available online at https://www.ncbi.nlm.nih.gov/books/NBK10981/, accessed January 31, 2024.

> Experimental data demonstrate that the psychological pain of social loss, such as the loss of a loved one, has neural correlates in the prefrontal cortex and the anterior cingulate cortex. Research . . . indicat[es] that when a person intends to suppress unwanted memories, his or her prefrontal cortex is involved in dampening activity in the hippocampus, a subcortical structure implicated in memory retrieval. Communicative intention between persons is signaled by the activation of two common brain regions . . . the same areas of the brain that activate when people are asked to consider the mental states of others. Emotion-induced memory gains and losses depend on a common neurobiological mechanism that can be manipulated by the pharmacological agent propranolol or by damage to the amygdala. A study establish[ed] that the brain's anterior cingulate cortex is implicated in monitoring the consequences of one's actions.[3]

The correlations between neural events, mental events, and the brain's neuroplasticity are clear and relatively uncontroversial findings of neuroscience. Thompson summarizes this data well: "Whenever we have a thought or a feeling, there is a corresponding firing pattern of these electrochemical charges along the neurons. . . . We do not experience anything without there being a corresponding neuron firing pattern."[4] He adds, "Many experiments have been done to identify the part of the brain that most often correlates with the activation of attention,"[5] and "Particular anatomical regions of the brain correlate with the mind's system of attention . . . different areas of the brain are associated with various forms of memory."[6] Due to these facts of neuroscience being relatively uncontroversial, I won't spend more time discussing neuroscience *per se*. My concern is how these findings relate to what we are and how we flourish.

FROM NEUROSCIENCE TO NEUROTHEOLOGY

Various interpretations have been offered to explain these findings. One interpretation is that of the neurotheologians. This interpretation then leads to an important conclusion, and from this conclusion follows the

---

3. Green et al., *In Search of the Soul*, 15–17.
4. Thompson, *Anatomy of the Soul*, 30.
5. Thompson, *Anatomy of the Soul*, 52.
6. Thompson, *Anatomy of the Soul*, 66.

neurotheologians' application to the question of how we flourish. What is the neurotheologians' interpretation of neuroscientific data?

Neurotheology's Interpretation: Our Thoughts Are Really Our Neurons[7]

A careful reading of Thompson and Wilder reveals they interpret the fact of correlation between the mental and the neural to mean that the mental ultimately *is* the neural. Thoughts are fundamentally neural firings. Feelings, in the final analysis, are in the brain. Choices are connections of synapses. In fact, they attribute all our behaviors, emotions, decisions, and even our very identity to brain activity. Here are some examples:

- "Remembering is *essentially* the process by which *neurons* increase their probability of firing together . . . essentially this is what Scripture points to when it speaks of 'the renewing of your *mind*' (Romans 12:2)."[8]

- "The *prefrontal cortex* contains neurons *responsible for* a range of complex, *conscious, intentional mental activity* . . . enabling us to: *discern* and *decide* . . . *distinguish* . . . create a mental sense of *expectations* . . . focus *attention* . . . [and] construct our *sense of morality* in the world."[9]

- "*Brain* functions . . . *determine* our *character*."[10]

---

7. Neurons are elongated cells that carry electrical impulses. They connect with other neurons via synapses—points of contact through which the electrical impulses travel.

8. Thompson, *Anatomy of the Soul*, 66–67 (emphasis added). "Essentially" is the key word here, which means *most fundamentally* or *ultimately*. For Thompson, remembering is, *at its core*, the firing of neurons.

9. Thompson, *Anatomy of the Soul*, 159, 162 (emphasis added). "Responsible for" again indicates that, in Thompson's view, the prefrontal cortex causes our mental activity. The prefrontal cortex (abbreviated as PFC) is the brain's highly developed frontal lobe. It is part of the cerebral cortex, the outer layer of the cerebrum (the top, outside, largest part of the brain, consisting of the brain's two hemispheres; what is usually shown in pictures of the brain). It is gray in color (as distinct from the brain's "white matter," which is large tracts of axons—the "tails" of neurons—that carry the electrical impulses of their associated neurons).

10. Wilder, *Renovated*, 6 (emphasis added). For Wilder, it is not our soul that determines whether we are honorable, just, or compassionate, but our brain.

- "The general tendency of our *hearts*—especially the *PFC [prefrontal cortex]*—is toward deceit and hiding the truth."[11]
- "Our *brain* creates and maintains a human *identity*."[12]
- "The *left hemisphere* [of the brain] sets me apart as '*me*.'"[13]

Neurotheology's Conclusion: Therefore, I Am Essentially a Brain

From this interpretation of the data, neurotheologians conclude the brain is the fundamental reality that stands under and unites all of our various experiences. In Thompson's words, "An integrating, oscillating wave of electrical activity is continuously moving back and forth across the entire brain. This wave may be one way that *the brain* brings together its disparate areas into a *convergent whole*, creating our overall sense of what we feel."[14] To clarify our brain is *fundamentally* what we are, rather than an immaterial "mind" (or "soul") he states, "The mind . . . is housed in your physical self and depends on your body to function"[15] and so, "no body, no mind."[16] For Thompson "the terms *brain* and *mind* . . . are . . . closely enough related to seem interchangeable."[17]

Wilder agrees: "The brain happens to contain a structure whose function is the integration of all internal states and external connections with others. The cingulate cortex (in the right brain) synchronizes

---

11. Wilder, *Renovated*, 170 (emphasis added). Here Wilder indicates that he believes the "heart" of a person is identical to the PFC, which again is designated by him as the cause of character traits (such as dishonesty).

12. Wilder, *Renovated*, 68 (emphasis added). Historically, the soul is what creates and maintains our identity (to be discussed in more detail in chapter 4).

13. Thompson, *Anatomy of the Soul*, 244 (emphasis added). Again, historically our individuation is a function of the soul, not the brain. See chapter 5 for more details.

14. Thompson, *Anatomy of the Soul*, 94 (emphasis added).

15. Thompson, *Anatomy of the Soul*, 29.

16. Thompson, *Anatomy of the Soul*, 31.

17. Thompson, *Anatomy of the Soul*, 9. Here Thompson does say that he does not use "mind" and "brain" to refer to identical concepts (though I believe he meant to say "identical entities," for he is discussing the nature of the mind and brain here, not his conceptions of them). However, though he promises to do so, I cannot find any place where he clarifies how he understands the mind and brain to be different. In chapter 3 I discuss nonreductive physicalism, the view that the mind emerges from the brain as smoke emerges from fire. This form of physicalism may explain what he means here and in a few other passages, though most of *Anatomy of the Soul* does seem to be written from the "reductive" physicalist perspective that the mind and brain are identical.

mental-energy states internally and externally."[18] Therefore, "The conscious attention of the slow track [of the brain] is usually what we mean by 'mind.'"[19]

From identifying the mind/soul with the brain, neurotheologians conclude that "I" am ultimately my brain. As Wilder observes, "When Dallas [Willard] describes our experience of the soul . . . he could hardly have described the cingulate in clearer terms."[20] (Wilder is severely misrepresenting Willard's view, as will be discussed further in chapters 7 and 9.) Wilder argues this is done as "a supraconscious brain process stays ahead of conscious thought. . . . The brain is constantly calculating the answer to *Who am I now?*"[21] Again, he uses "soul" and "brain" interchangeably when he states, "The soul integrates our identities and directs the energy of everything it means to be human. The brain can create this integration using the cingulate cortex."[22]

Thompson goes on to explain how he believes the brain makes us us: "The *brain* is constantly scanning the internal and external landscape, *comparing* the present to the past in order *to prepare* the body for the future."[23] He adds, "The heart—our deepest emotional/cognitive/conscious/unconscious *self*—is manifested most profoundly at the level of the prefrontal cortex."[24] As a psychiatrist, this is what Thompson says he most wants his patients to understand, for "a better working understanding of the structures and function of the brain . . . gives [patients] a greater appreciation of what makes them *uniquely human*."[25] He states, "The prefrontal cortex (PFC), along with our language centers, is the part of *our neurological system that sets us apart* from all of God's other created beings."[26]

---

18. Wilder, *Renovated*, 85. The cingulate cortex is the inside surface area between the two sides of the cerebrum, not visible in typical pictures of the brain.

19. Wilder, *Renovated*, 86.

20. Wilder, *Renovated*, 85.

21. Wilder, *Renovated*, 36 (emphasis in original).

22. Wilder, *Renovated*, 89.

23. Thompson, *Anatomy of the Soul*, 185.

24. Thompson, *Anatomy of the Soul*, 169 (emphasis added).

25. Thompson, *Anatomy of the Soul*, 31 (emphasis added). Again, historically it is the soul, which includes the image of God, that makes people uniquely human. See chapters 2 and 5 for a further discussion of the historic understanding.

26. Thompson, *Anatomy of the Soul*, 157 (emphasis added).

Neurotheology's Application: Neuroscience Is the Key to My Flourishing

From this conclusion that I am ultimately my brain, the neurotheologians' application follows: the better we understand neuroscience, the better we will understand what we are and therefore how we flourish. Neuroscience is the key to understanding the full reality of our experiences and the pathway to healthier emotions and truer beliefs, which lead to proper desires and better choices.

For instance, Wilder's book is a result of the 2012 "Heart and Soul Conference" that he helped coordinate to "explore, for the first time, how *the brain* learns Christlike character."[27] A better understanding of neuroscience is, as the title of chapter 9 in *Renovated* puts it, "The Science of Spiritual Maturity."[28] Thompson concurs: "New discoveries in neuroscience and related fields offer clues as to how you can develop these attributes [of] joy, goodness, courage, generosity, kindness, and faithfulness."[29] As the title of his book illustrates, by understanding the *Anatomy of the Soul* (the neurophysiology of the brain), we will find "surprising connections between neuroscience and spiritual practices that can transform life and relationships."

NEUROTHEOLOGIANS ARE PHYSICALISTS

The underlying idea being promoted by these neurotheologians, knowingly or unknowingly, is a philosophical position known as *physicalism*.[30] Physicalists believe that only physical things exist—only physical things are real.[31]

We inherited physicalism from the Enlightenment,[32] and it now shapes every aspect of Western culture. It permeates how stories are told

---

27. Wilder, *Renovated*, 2 (emphasis added).
28. Wilder, *Renovated*, 145–70.
29. Thompson, *Anatomy of the Soul*, xvii.
30. The same idea is often referred to as materialism. Throughout I have chosen to use the term "physicalism." For the history of these terms, as well as how they may be used in slightly different ways, see the *Stanford Encyclopedia of Philosophy* under "Physicalism," https://plato.stanford.edu/entries/physicalism/, accessed August 29, 2023.
31. Or at least physical things are the basis of all reality. This distinction will be developed in subsequent chapters.
32. The period of Western intellectual history that generally took root in the 18th century.

in books and movies, how current events are interpreted, and how school curricula are designed. We bump into it in conversations with children, friends, and colleagues. For instance, your son may come home and tell you that his high school science teacher said, "The universe is all there is, was or ever will be" (quoting the famous physicalist Carl Sagan). Or during the Easter season, you may be talking with a co-worker about the resurrection and she says, "There is no way someone rises from the dead. Miracles can't happen. Even if scientists can't explain it now, some day they will." Or perhaps your neighbor buys his fifth sports car (and is now planning to build a bigger garage) and says to you, "You only go around once, so the one who dies with the most toys wins."

Again, knowingly or unknowingly, following from this philosophical commitment is Wilder and Thompson's *anthropology* (understanding of what persons are): we are, like the rest of reality, fundamentally physical beings. In his wonderful survey of the major *worldviews* (overarching philosophies of life), James Sire encapsulates this physicalist anthropology: "Human beings are complex 'machines'; personality is an interrelation of chemical and physical properties we do not yet fully understand."[33] Or, to quote a famous physicalist, "'You,' your joys and your sorrows, your memories and your ambitions, your sense of identity and free will, are in fact no more than the behavior of a vast assembly of nerve cells and their associated molecules."[34]

This anthropology is prevalent in our popular culture as well. For instance, you might see a documentary about how people are being frozen just before death so that when a cure for their disease is found, they can be unfrozen and cured, based on the assumption that we are no more than complicated machines that can be "ice boxed" indefinitely until repairs can be made. (This is known as cryogenics. It was the basis of the classic Mel Gibson movie *Forever Young*.) Or suppose that a wealthy alumnus of your university makes a large gift to build a new library but passes away before the library is built. When it is completed, the university president names the library after this generous donor, saying, "Mrs. Jones will live on through this library's influence in the lives of countless students."

But Thompson, Wilder, and other neurotheologians are followers of Christ. They certainly believe some things exist beyond what can be seen—such as God. Therefore, perhaps interpreting them to be saying

---

33. Sire, *The Universe Next Door*, 56. Sire uses the term "naturalism" as a synonym for physicalism.

34. Crick, *The Astonishing Hypothesis*, 3.

that we should reduce immaterial things like thoughts and ultimately souls to material things like neural firings and ultimately brains is unfair. Perhaps they are not physicalists, making an ontological[35] claim that mental events are ultimately brain events (and more broadly that our souls/immaterial dimensions are ultimately our bodies/material dimensions). I want to be as charitable as possible. So two other interpretations of what they write must be considered.

Are Neurotheologians Simply Using Complementary Explanations?

First, perhaps they are offering only one of two complementary explanations of our will, choices, and so on, while also maintaining that the existence of the mind or soul as an equally true explanation.

We have many examples, including examples in Scripture, of complementary explanations providing different but equally true descriptions of an event. The laws of astronomy and the claim that Christ was "before all things, and in him all things hold together" (Col 1:17) are both true, but from different perspectives. God's parting of the Red Sea for his purposes (Exod 14) and an earthquake or other natural phenomenon occurring as the physical cause are complementary explanations from theological and physical perspectives. Perhaps this is what Wilder and Thompson are doing—giving explanations of our mental life from the perspective of neuroscience, while not discounting equally true accounts of our mental life as activities of our immaterial dimension, i.e., our soul.

Though possible, this seems improbable. Nowhere do they say that this is what they are doing. We could read this idea into their writings, but this would be putting words in their mouths. It seems more reasonable to take them at their word (for instance, the passages from their books cited above) and conclude that they view all our experiences as ultimately physical. Furthermore, it seems more charitable to assume they mean what they say, rather than something else they inadvertently failed to communicate or (worse yet) chose not to communicate.

This thesis that Wilder and Thompson embrace complementary explanations of events seems even more improbable when we move from the level of specific passages in their books to their overall thesis.

---

35. Ontology is the study of what is—the nature of being, from the Greek *ontos*, "being" or "that which is."

Their desire is to help us understand what we are and therefore how we flourish. Given the importance of this project, and the importance of first having an accurate understanding of what we are, it is fair to assume that *if* Thompson and Wilder believe theirs is only *one* explanation of what we are (the material explanation), they would make it a point to remind readers that there is another, equally true immaterial explanation of what we are—a soul. This would seem especially important as they move to application—showing how we can grow in Christ. But nowhere do they indicate they are only telling us half the story. Rather, the impression is that they are offering a complete explanation of what we are and therefore how we flourish. This is further reason to discount the complementary thesis.

Beyond their writings, there is an additional reason to believe that both Thompson and Wilder are in fact physicalists. Thompson writes, "I have chosen to provide the scientific data behind my ideas in a bibliography of books that have influenced me."[36] The first section of his bibliography lists sources important to him in understanding what we are. It is entitled "Science and the Mind/Relationship Matrix." There he lists the works of many physicalists, including Daniel Siegel, author of *The Developing Mind*, which he identifies as a "landmark book."[37] Missing from this section of his bibliography are any neuroscientists, theologians, or philosophers who study this issue from a non-physicalist perspective. From this information, it is clear that he has been deeply influenced by physicalists who explicitly reduce the soul to the body and the mind to the brain.

The same is true of Wilder. He quotes approvingly physicalists such as Siegel, who "would call my awareness 'mindsight.'"[38] This physicalist idea is a central aspect of Wilder's book *Renovated*. He also cites Thompson, who spoke at the 2012 Heart and Soul Conference, the themes of which form the basis for *Renovated*. Wilder writes, "In his work on mindfulness, Dr. Curt Thompson . . . combines neuroscience and spiritual practices with special attention to the effects of mindfulness on relationships."[39] Given that Thompson bases his ideas on physicalists, and that Wilder is basing his ideas on Thompson and Siegel among others, it is fair to conclude that Wilder's thought is based on physicalist assumptions as well.

36. Thompson, *Anatomy of the Soul*, 9.

37. Thompson, *Anatomy of the Soul*, 6, citing Siegel, *The Developing Mind*.

38. Wilder, *Renovated*, 33, citing Siegel, *Mindsight: The New Science of Personal Transformation*.

39. Wilder, *Renovated*, 38.

This is even more reasonable considering whom Wilder studied under in seminary. Two of the leading Christian physicalists are at Fuller Seminary, where Wilder earned his MA and PhD. The well-known scholar Joel Green, a senior professor of New Testament interpretation, advocates what is best described as a "neuroscientific hermeneutic" in interpreting the Scriptures. This is roughly the notion that in approaching the Bible, one should try to show that its teachings on human nature are consistent with the idea that we are purely physical beings, which he believes is justified by neuroscience.[40] Also at Fuller is philosopher Nancey Murphy, another vocal Christian physicalist. She unequivocally states, "We need not postulate the existence of an entity such as a soul or mind in order to explain life and consciousness."[41] It is reasonable to assume that Murphy and Green had a deep influence on Wilder's thinking.[42] Therefore, it is very unlikely that either Wilder or Thompson are providing only a complementary, rather than a physicalist, explanation of what we are.

Are Neurotheologians Just Imprecise?

Or perhaps Thompson and Wilder truly believe we are enduring souls and are just imprecise when they use words such as *essentially*, *responsible for*, *creating*, and *determining* to refer to the brain's role in thinking, believing, desiring, choosing, and being a human person. Might they actually mean, but not communicate clearly, that the soul *uses* the brain in these activities?

The responses above also apply here. But even if the impression that Wilder and Thompson are physicalists is merely the result of their imprecise use of these and similar words, it is troubling that they didn't clarify

---

40. Green, *Body, Soul, and Human Life*, 21–34, see especially 28–29.

41. Murphy, "Human Nature: Historical, Scientific and Religious Issues," 18. Stuart L. Palmer connects the influence of Siegel to the physicalism of Nancey Murphy: "Murphy, it appears, is significantly influenced in the field of interpersonal neurobiology. See, e.g., Daniel J. Siegel, *The Developing Mind*." Palmer, *In Search of the Soul*, 201, note 16. I'll discuss Murphy's nonreductive form of physicalism in chapter 3.

42. Wilder had hoped to further develop his understanding of what we are and how we flourish by studying with Dallas Willard, but Willard passed away just before he had this opportunity. He writes, "The Biola [University] Center for Christian Thought had received a grant for a fellowship year integrating neuroscience with theology under Dallas Willard. With Dallas's encouragement, I applied, but Dallas did not make it there." Wilder, *Renovated*, 29.

how they were using these words, to make sure they didn't give anyone the impression we are fundamentally material. Words matter. How we talk about ourselves matters. So even if the problem is "only" their use of imprecise words, this in itself should be deeply troubling.

Therefore, whether Thompson and Wilder are actually physicalists or just write as if they are,[43] it is imperative for us to evaluate this anthropology, along with its implications for (or its impediments to) our flourishing. As Proverbs 18:17 says, "The first to state his case seems right until another comes and cross-examines him." This cross-examination is crucial, because wrong beliefs about ourselves that are not identified, clarified, and explicitly evaluated are the most pernicious types of beliefs. It is hard to correct wrong beliefs of which we are not even aware! The way to begin is by exploring additional information we have about our nature.

WHAT ELSE DO WE KNOW ABOUT WHAT WE ARE?

Christians know that many things exist beyond the realm of sensory perception, such as God, angels, and moral values. Therefore, the advances in neuroscience do not rule out the possibility that we are immaterial persons, if we have other good reasons to believe that we have a soul as well as a body. To seek additional knowledge of what we are we can look beyond science to God's special revelation: truths God has revealed in Scripture. This is the domain of theology. We can also look to other fields, in addition to science, that study God's general revelation (truths he has revealed in the created order). Philosophy, my primary field of study, also researches in much detail God's general revelation about what we are.

Philosophy comes from two Greek words: *phileō* (love) and *sophia* (wisdom, or skillful living). It is the love of wisdom about reality and thus how to live well and flourish. Proverbs tells us all to be philosophers: "Seek her [wisdom] as silver, and search for her as for hidden treasure." (Prov 2:4) The apostle Paul studied philosophy to the point of being able to quote pagan philosophers such as Epimenides and Artus in Acts 17:28, "as even some of your own poets have said, 'For we also are his children.'"[44] In 1 Peter 3:15, we are told we should "always [be] ready to

---

43. In chapter 3, I will consider another form of physicalism, which may be the variety Wilder and Thompson actually believe. The point here is that the view they are expressing is indeed physicalism, regardless of which variation they ultimately endorse and whether they do so intentionally or unintentionally.

44. Philosophers were often referred to as poets in this era.

make a defense to everyone who asks you to give an account for the hope that is in you." The word used for "defense" (*apologia*) meant providing arguments for your position and against contrary positions in a court of law. Doing so requires logic and reason, cornerstones of philosophy.

Philosophy has played a crucial role at important junctures in church history. For instance, in the fourth century A.D. there was much discussion about how best to understand the nature of the Trinity. The solution was found in the philosophical concepts of substance or essence (*ousia*) and person (*hypostasis*), which were important ideas in Greek philosophy. The Nicene Creed (325) states the Son is "the same substance with the Father" (*homoousios*—having the same substance/essence). Later in the same century, the Council of Constantinople (381) similarly summarized biblical teaching that Jesus is "true God from true God, begotten, not made, of the same substance (*homoousion*) with the Father." Yet the Father, Son, and Spirit are different persons, as Scripture teaches. Therefore, the Athanasian Creed (fifth century) said we must "worship one God in Trinity, and Trinity in Unity; neither confounding the Persons (*hypostases*), nor dividing the Essence (*ousia*). For there is one Person (*hypostasis*) of the Father; another of the Son; and another of the Holy Ghost."

In this way, philosophy helped provide a better understanding of the Trinity by clarifying concepts such as what a "nature" is and what a "person" is. This use of philosophy in making distinctions, clarifying concepts, and drawing out logical implications about what a thing is can also help us better understand what we are.

In sum, we must integrate *all* we know about what we are from science *and* theology *and* philosophy. In doing so we will we have a more accurate understanding of our nature, necessary to better understand how we flourish. Do theology and philosophy suggest a different interpretation of the data of neuroscience? As I alluded to several times already, I believe they do.

2

The Bible and the Soul

Everything in the Scriptures is God's Word. All of it is useful for teaching and helping people and for correcting them and showing them how to live.
—2 Tim 3:16 (CEV)

What is the human? Yes, that is a most important question, to which the biblical revelation gives the best answer.
—Millard Erickson[1]

> **CHAPTER SUMMARY**
>
> Beginning with an affirmation of biblical revelation as a means to understand reality, including the reality of what we are, this chapter evaluates the biblical texts related to anthropology. Thompson and Wilder's references to relevant biblical texts are first discussed and found inadequate. A more robust engagement with the Scriptures uncovers a number of truths about what we are. It affirms the importance of our bodies as a central feature of our constitution. Yet it also affirms the reality and importance

---

1. Erickson, *Christian Theology*, 495.

> of our souls, which reflect the image of God. Furthermore, we find the body and soul to be deeply united, which is how God created us to be, desires us to be, and will restore us to be at the final resurrection. However, this fact also indicates that we are ultimately a soul, since Scripture teaches that we live on after our bodies die. Therefore, we must most fundamentally be a soul, and not a body or a composite of soul and body. The chapter concludes with a brief survey of how this understanding of the biblical teaching on what we are has been affirmed by believers throughout the centuries.

UNIVERSITIES WERE FOUNDED TO foster conversation among various fields of study, in order to best discover truth. Their very name, *universities*, reflects the foundational idea that there is a unity to knowledge, discovered in each discipline and refined through the integration of the findings in one field of study with the findings of other fields, into one robust whole.

Historically, theology was included as an academic discipline, because it was understood to provide knowledge of reality, as do all other academic disciplines. Therefore, every other field had to interact with what is known from the Scriptures. Unfortunately, amidst the Enlightenment's physicalism, theology was no longer seen as providing knowledge of reality. Therefore, there was no longer any sense that theology offered truths that had to be integrated with discoveries in other fields.[2] This separation of theology from other areas of knowledge has now flowed from our universities to every sector of society.

---

2. As a result, theology departments began to disappear in universities. They were replaced by religion departments, which study how people express their religious beliefs, regardless of whether the object of belief exists. For a more thorough assessment of the removal of theology as an academic discipline in pluralistic universities, and of the resulting disintegration of higher education, see Klassen and Zimmermann, *The Passionate Intellect*, Part 2: "The Story of Humanism from Its Holistic Medieval Beginnings to Postmodern Fragmentation," 45–119. See also Ruben, *The Making of the Modern University*.

As Christians, we must reject this "dis-integration." Scripture truly is "God-breathed" and provides us with knowledge on the topics it addresses. In the case of anthropology, biblical scholars have been discussing for centuries what the Scriptures say about our nature. Biblical texts in both the Old and New Testament provide a treasure trove of information. We must integrate this knowledge with what is known from neuroscience.

NEUROTHEOLOGY'S BIBLICAL ANTHROPOLOGY

Thompson and Wilder seem to agree with the idea that we should integrate what we know from Scripture into our understanding of ourselves. In various places, they offer biblical passages to illustrate their points.

In *Renovated*, Wilder identifies various biblical texts that focus on our need for loving relationships with others—"attachment love," as he calls it.[3] His thesis is that we must be rightly related to God first and then to others in order to live flourishing lives.

Yet this first requires something to stand *in* these relationships of attachment love. As Wilder rightly notes, "*Attachment* or *bonding* is language that comes from gluing things together."[4] So what is attached? He interprets Scripture to be saying it is the brain that enters relationships: "the brain has attachment love."[5]

Thompson starts with the assumption that the terms "mind" and "soul" are important anthropological terms in Scripture. In chapter 3 of his book, titled "Love the Lord Your God with All Your . . . Mind," he identifies the central passages that refer to our minds. He writes, "Jesus also speaks of our minds—perhaps most notably he tells us to 'love the Lord your God with all your heart and with all your soul and with all your mind' (Matt 22:27)."[6] Thompson then cites three other important passages that refer to the mind: "We have the mind of Christ" (1 Cor 2:16); "The mind controlled by the sinful nature is death, but the mind controlled by the Spirit is life and peace" (Rom 8:6); and "Do not conform

---

3. These passages are listed in *Renovated*'s endnotes, on pages 213–20. They include, among others, Eph 3:17, Rev 2:4, 1 Thess 3:5–7, and Col 3:9–10.

4. Wilder, *Renovated*, 113.

5. Wilder, *Renovated*, 75. He repeats later on page 111, "Relational-brain skills . . . are the activities of a smoothly-running fast track in the brain."

6. Thompson, *Anatomy of the Soul*, 28.

to the pattern of this world, but be transformed by the renewing of your mind" (Rom 12:2).[7]

Yet Thompson doesn't believe there is much to understand theologically concerning these biblical references to "mind" or "soul." Consistent with his physicalist anthropology, he immediately turns to neuroscience: "While God doesn't spell out for us exactly how the mind works, scientific research provides intriguing clues."[8] He then reads all biblical texts that mention the soul, mind or specific mental activities as referring to the brain or to specific neural properties studied by neuroscience. For instance, after citing a number of biblical texts concerning the importance of memory, he concludes that "remembering is essentially the process by which neurons increase their probability of firing together."[9] In other words, memories are in the brain and are studied by neuroscience. The same is true of emotions: "Let's define what we mean by emotion . . . as with all areas of brain function."[10]

Thompson's understanding of biblical references to the mind as actually references to the brain appear in his discussions of the mind of Christ as well. He observes, "Jesus' mind, I suggest, reflects the most integrated prefrontal cortex of any human of any time."[11] This "mind of Christ" is seen, for instance, in Jesus' resisting temptation: "Essentially, Satan suggests that Jesus use his gifts as coping strategies in the face of anxiety. Doing so, however, would have cut Jesus off from various parts of his own mind—his own prefrontal cortex."[12] This physicalist explanation of biblical texts continues throughout later chapters as well.[13] Of the role of neuroscience in helping us understand what "God doesn't spell out" about our nature in Scripture he summarizes, "Neuroscience acts like a magnifying glass, enabling us to see detail about *the human condition that we might otherwise overlook*."[14]

---

7. Thompson, *Anatomy of the Soul*, 28.
8. Thompson, *Anatomy of the Soul*, 67.
9. Thompson, *Anatomy of the Soul*, 67.
10. Thompson, *Anatomy of the Soul*, 91.
11. Thompson, *Anatomy of the Soul*, 180 (emphasis in original).
12. Thompson, *Anatomy of the Soul*, 181.
13. For instance, see chapter 10, "Neuroscience: Sin and Redemption," 183–203 and chapter 13, "The Mind and Community: The Brain on Love, Mercy, and Justice," 235–55 (again equating "brain" and "mind").
14. Thompson, *Anatomy of the Soul*, 205 (emphasis added).

A MORE ADEQUATE BIBLICAL ANTHROPOLOGY[15]

But God *does* spell out much more about what we are in both the Old and New Testaments. While it is true that the main focus of Scripture is to trace God's redemptive plan, and so it does not focus on the question of anthropology *per se*, many texts do imply a much more detailed view of what we are than Wilder or Thompson indicate. The Scriptures affirm the importance of the body, but also affirms that we have a mind, or soul. And although the soul and body are deeply united, the fact that we will, for a time, live apart from our bodies after we die and before the final resurrection demonstrates that we are ultimately a soul that has a body.

Our Body Is Crucially Important

Scripture clearly assumes that our bodies are vitally important to what we are. First Corinthians 6:19 tells us, "Your body is a temple of the Holy Spirit within you, whom you have from God." As Thompson rightly observes, "Paul describes our bodies as the temples of the Holy Spirit, so clearly they're involved in our deepest spiritual experiences."[16] The value of the body is ultimately affirmed by God himself becoming incarnate: "And the Word became flesh, and dwelt among us" (John 1:14). Finally, believers look forward to the final resurrection where we will receive a new body (2 Cor 5:1–5), indicating the everlasting value of our physicality.[17]

We Also Have a Soul[18]

As important as the body is, the common refrain throughout Scripture is that we also have an immaterial dimension, typically referred to as our mind, or in the biblical literature, our soul.

---

15. Due to space constraints, I will offer only a summary of the important biblical texts. For a much fuller discussion, see Cooper, *Body, Soul, and Life Everlasting*. This is an extremely thorough and nuanced discussion of all the biblical texts touching on anthropology, which I will rely on heavily in this section. For a briefer, more accessible treatment, see Moreland, *The Soul: How We Know It's Real and Why It Matters*, 40–73.

16. Thompson, *Anatomy of the Soul*, 29.

17. Other passages discussing the reality and importance of the body include Rom 6:11–13, 19; 12:1; 1 Cor 9:24–27; 1 Tim 4:7–8.

18. I will develop a much fuller description of the soul in chapter 5 and its relation to the body in chapter 6. Here I am using the term simply to refer to our immaterial dimension, however it is construed.

Genesis 2:7 speaks of God creating us as body *and* soul: "Then the Lord God formed the man of dust from the ground, and breathed into his nostrils the breath of life; and the man became a living person." When we were just material ("the dust of the ground"), we were not yet persons. Not until God added "the breath of life" (the Hebrew word used is *neshama*—the human spirit[19]) did we become persons. We are more than a body. We also have an immaterial dimension—a spirit or a soul.

The other Hebrew term often translated as "soul" is *nephesh*.[20] For instance, it appears in Psalm 30:3: "Lord, You have brought up my soul from Sheol; You have kept me alive, that I would not go down to the pit." This idea of our soul also appears in Genesis 35:18: "And it came about, as her soul was departing (for she died), that she named him Ben-oni; but his father called him Benjamin."

The fact that we have a soul, in addition to a body, is also referred to in the New Testament. Jesus cautions us to "fear Him who is able to destroy both soul and body" (Matt 10:28).[21] He clearly has our material and immaterial dimensions in mind. Elsewhere he warns, "For what good will it do a person if he gains the whole world, but forfeits his soul? Or what will a person give in exchange for his soul?" (Matt 16:26). Jesus draws a clear distinction between our bodily life, which can gain all the world has to offer, and the life of our soul, which can lose all that eternity has to offer. The soul is also referred to in passages that discuss our salvation, such as 1 Peter 2:25: "For you were continually straying like sheep, but now you have returned to the Shepherd and Guardian of your souls."

Christian physicalist Joel Green claims that in these passages the reference to the soul is a figure of speech called a synecdoche, i.e., when

---

19. This term is also used of other living creatures and is parallel to the term for God's spirit, *ruach*. Cooper summarizes: "*Ruach* . . . refers to the spirit of God more frequently than to the human spirit. When indicating the breath of a living creature *ruach* is often parallel to another important term, *neshama*, 'the breath of life' (Gen 2:7)." Cooper, *Body, Soul, and Life Everlasting*, 39.

20. *Nephesh* occurs 754 times in the Old Testament. It usually refers to humans, but sometimes it refers to God or animals. It has three meanings. It can mean the soul of a person that can live apart from the body. It can also mean more generally a life principle. Finally, it is sometimes used figuratively. See Brown et al., *A Hebrew and English Lexicon*, 220. The passages I cite seem clearly to be using the term to refer to the soul of a person.

21. The Greek word here translated "soul" is *psychē*. This is one of two Greek words used in the New Testament to refer to our immaterial dimension (the other being *pneuma* [spirit], discussed below). Cooper discusses the New Testament use of *psychē* in Matthew 10:28 in *Body, Soul, and Life Everlasting*, 117–19. He discusses its use in Revelation 6:9–11 on pages 115–17.

a part is named in order to refer to the whole thing that contains that part.[22] Thus, he proposes that these passages actually refer to the whole person, who is physical. But this response fails. To see why, here is a good example of this figure of speech: "All hands on deck!" Clearly, this expression is talking about entire people, commanding them to be on deck. But note that in cases like this synecdoche of part for whole, they still affirm the reality of the part! Even if whole persons are in mind, there are still such things as hands. Similarly, even if the whole person (body and soul) *is* in mind in these biblical texts,[23] the figure of speech still affirms the reality of the soul!

Understanding that we have a soul is so important because the soul is what reflects the image of God (the *imago Dei*) described in Genesis 1:26: "Then God said, 'Let Us make mankind in Our image, according to Our likeness.'"[24] Whatever else can be said about the *imago Dei*, it is not material. God is an immaterial person, so his attributes must also be immaterial, including his image. He shares some of his attributes with us (these are known as God's *communicable* attributes), including his image. Yet in order to do so, we too must have an immaterial dimension in order to exemplify these attributes. Understanding this helps us understand our nature, for as Alvin Plantinga says, "God is the premier person . . . and the properties most important for an understanding of our personhood are properties we share with him."[25]

We may conclude, on the basis of this biblical data, that we also have an immaterial dimension. We have a soul.

---

22. See for instance Green, *Body, Soul, and Human Life*, 151.

23. I don't grant that the whole person, body and soul, is always the referent in the texts cited above.

24. I am taking the image of God to be an ontological reality—something that actually exists. Others argue that the *imago Dei* refers to our role representing God in creation. Still others say it is relational, signifying the relationships we have with God and others that are like God's relationship to his creation. Yet both of these last two interpretations implicitly assume that the *imago Dei* is first an ontological reality. We can represent God only if we first possess certain attributes in common with him that form the basis of our representation. And the type of relationships a thing can enter is dependent on the nature of that thing which exists. So the ontological reality of the *imago Dei* is fundamental and grounds its representational and relational properties. For a much fuller discussion of the image of God and the implications for physicalism, see Moreland, *The Recalcitrant Imago Dei*. For a fuller theological treatment of the image of God, see Hughes, *The True Image: The Origin and Destiny of Man in Christ*. More will be said of the image of God in chapter 5.

25. Plantinga, "Advice," 264–65.

The Body and Soul Are Deeply United

Furthermore, Scripture indicates that these two dimensions—body and soul—are deeply united to one another. The Old Testament emphasizes this unity throughout. In Ezekiel 37:5–6 we read, "This is what the Lord God says to these bones: 'Behold, I am going to make breath enter you so that you may come to life. And I will attach tendons to you, make flesh grow back on you, cover you with skin, and put breath in you so that you may come to life; and you will know that I am the Lord.'" Both body and soul, deeply intertwined, are essential for life on this earth. This unity is assumed in the New Testament as well. For instance, the disciples assumed persons to be a unity of body and soul, and so when Jesus appeared after the resurrection, they were shocked by what they thought they saw—a spirit without a body: "But they were startled and frightened, and thought that they were looking at a spirit" (Luke 24:37).

This deep unity makes sense of the final resurrection, when we will again be whole and will live forevermore embodied, as we were always meant to be. "For our citizenship is in heaven, from which we also eagerly wait for a Savior, the Lord Jesus Christ; who will transform the body of our lowly condition into conformity with his glorious body" (Phil 3:20–21).[26]

The Christian faith is unique in emphasizing the importance of the body for full human functioning. Accordingly, N. T. Wright expresses the biblical depiction of our bodily resurrection as "life after life after death."[27] We die and live without our bodies for a period of time in the presence of God. At the end of the age, our bodies are then resurrected and reunited with our souls. In God's redeemed creation, where all things are put right, we live in our new and imperishable bodies forever. This indicates that it is good for us to be embodied, and is in fact our "natural" state—how we were created to exist.[28]

It is a common error to deny the deep unity between our material and immaterial dimensions. This was the error committed by the

26. Also, for instance, Rom 8:23: "We ourselves, having the first fruits of the Spirit, even we ourselves groan within ourselves, waiting eagerly for our adoption as sons and daughters, the redemption of our body"; and 1 Cor 15:42–43: "So also is the resurrection of the dead. It is sown a perishable body, it is raised an imperishable body; it is sown in dishonor, it is raised in glory; it is sown in weakness, it is raised in power." There is an order to this resurrection: Christ was raised as the first of the harvest; then all who belong to Christ will be raised when he returns (1 Cor 15:23).

27. N. T. Wright, *Surprised by Hope*, 151.

28. See 1 Cor 15:42–57. For a helpful discussion of this text, see Morris, *The First Epistle of Paul to the Corinthians*, 221–30.

Gnostics in the first and second centuries, as we saw in the introduction to this book. This error has led to many excesses and deficiencies throughout church history, including denying the reality of the body's needs or the body's value in flourishing and spiritual formation. In ministry, it has led many to focus exclusively on people's "spiritual" needs and ignore their physical needs, which we should also seek to meet. This error has also led many to suppose that "spiritual" activities such as Bible study and full-time Christian ministry are better than "mundane"[29] activities such as the study of "secular" topics or working in "secular" professions.[30]

These errors are often justified by erroneously assuming that our bodies, and more broadly the physical dimension of reality, are either not valuable at all or at least less valuable than the spiritual realm. A robust, biblical understanding of the reality and goodness of the body and the deep, natural, and therefore valuable relationship between our material and immaterial dimensions goes a long way to correct these errors.

Yet Ultimately We Are a Soul That *Has* a Body

As problematic as denying the unity of our material and immaterial dimensions is, it is equally problematic to make the opposite error: denying that the soul and body are distinct. Yet this error is equally prevalent—and perhaps now even more so.

This error is certainly understandable. Biblical references, especially in the Old Testament, seem to emphasize that the human person is a unity. Hebrew words often translated "soul," "spirit," "flesh," "inner parts," and "heart" can often be interpreted as referring to the whole person.[31]

From this, some assume that this Hebraic sense of unity must be an *ontological* unity—that in reality, or in the core of our being, we are ultimately one thing. This is often argued on the grounds that the idea of an ontological dichotomy between body and soul is a Greek idea, not a Hebrew one. Rather, some claim, the Hebrew mind understood persons to be an ontological unity. Therefore, any notion of duality must be rejected as misreading Scripture through the lens of Greek thought.

---

29. The word "mundane" derives from the Latin word *mundus* which means "worldly" (as opposed to heavenly or spiritual).

30. For more on a proper theology of vocation, see Ryken, *Redeeming the Time*; Crouch, *Culture Making*.

31. For more on the use of these words in the Old Testament and the implications, see Cooper, *Body, Soul, and Life Everlasting*, 38–51.

However, both Greek and Hebrew thought actually had a range of views of the human person, from ontological duality to ontological unity. As Cooper summarizes,

> Although the influence of Hellenism is difficult to deny, it does not appear to have radically redirected or transformed the ethos of most Jewish anthropology. Affirming a dichotomy of body and soul . . . does not necessarily contradict its holistic emphasis on human life and seems wholly compatible with Old Testament anthropology.[32]

Since these Old Testament passages don't necessarily indicate that a person is an ontological unity, they may be referring to another type of unity: *functional* unity. In this case, our material and immaterial dimensions are understood as functioning in unison (each deeply affecting the other). Yet functional unity does not entail ontological unity. Consider a car. It is a functional unity in that all of its parts function together according to what the car is designed to do. But it is also ontologically diverse, since numerous different components make up the car. Such may be the case with us: we may be a functional unity while still being two different things—an ontologically distinct body and a soul.[33] This would explain equally well the Old Testament references to human beings as a unity.

Is there any indication whether these biblical passages should be interpreted to mean we are a *functional* unity or an *ontological* unity? A basic principle of biblical interpretation is to allow Scripture to interpret Scripture. In other words, we must allow other passages with more clarity to help us interpret the passages that are less clear and, on their own, ambiguous, as in this case of whether the unity referred to in some Old Testament texts is ontological or functional.

As briefly noted earlier, a number of biblical passages seem to indicate that we can live on after our bodies die. Therefore, if *we* continue to exist without our bodies, it follows that *we* are ultimately that which continues—our immaterial dimension, or our soul that presently has a body. In this case, we must be an ontological duality of soul and body. A more thorough evaluation of biblical texts shows that this is clearly the case.

---

32. Cooper, *Body, Soul and Life Everlasting*, 86.

33. Continuing to assume the only type of unity is ontological is a logical fallacy known as a false dichotomy. It assumes that there must be either ontological unity of the person or no unity at all, when functional unity is a viable third option. See chapter 8 for more.

A survey of all Old Testament texts finds that "in Hebrew theology, one can see over time the gradual affirmation of the reality of an afterlife (Gen 37:35; Job 3:13; Song 3:1–4), and this seems to require some distinction between the soul (or person) and the body—as we find in Ecclesiastes 12:6–7."[34] N. T. Wright states, "Any Jew who believed in resurrection, from Daniel to the Pharisees and beyond, naturally believed also in an intermediate state in which some kind of personal identity was guaranteed between physical death and the physical re-embodiment of resurrection."[35]

This idea that the soul continues when the body dies becomes even more clear in the New Testament.[36] In Matthew 17:1–13, we read of Jesus going to a mountain and meeting Moses and Elijah, who had long since passed from this earth. Then in Matthew 22:23–34, Jesus sides with the Pharisees against the Sadducees in affirming that Abraham, Isaac, and Jacob are still alive, though their bodies were buried long ago.[37] Paul says precisely the same thing in Acts 23:6–9.

Perhaps most direct and succinct is Jesus' promise to the thief dying on the cross next to him. Jesus assured him, "Today you will be with Me in paradise" (Luke 23:43). This is not a parable or an analogy. It is not an expression of how things will (erroneously) seem to the thief. Rather, it is a straightforward statement of what will happen very shortly. Though their bodies (including their brains) will soon be dead, *they* will be very much alive and together, enjoying paradise. They are not to be identified with their bodies (the material dimension) but with their souls (their immaterial dimension).

Soon after this, Jesus dies: "And Jesus, crying out with a loud voice, said, 'Father, into Your hands I entrust My spirit.' And having said this, he died" (Luke 23:46). He commits to God what he essentially is—his spirit[38]—to live on after his body dies. We further see in 1 Peter 3:18–19

---

34. Goetz and Taliaferro, *Brief History of the Soul*, 31. "Remember your Creator before the silver cord is broken . . . then the dust will return to the earth as it was, and the spirit will return to God who gave it" (Eccl 12:6–7).

35. Wright, *The Resurrection of the Son of God*, 164.

36. Therefore, even if the Old Testament texts were taken as providing a *prima facie* rationale to understand us as an ontological unity, the further revelation provided in the New Testament shows that the best understanding of these Old Testament texts is, in fact, describing us as an ontological duality.

37. See France, *The Gospel According to Matthew*, 316–18.

38. The Greek word used here is *pneuma*. This is the other word used in the New Testament to refer to our immaterial dimension. Cooper states that when the authors of the Gospels use this word, they mean "personal spirit without a fleshly body." He

that Jesus did not die when his body died.[39] Whatever the best interpretation of the details of this passage may be, it is clear that Jesus continued to have a conscious existence while his body lay in the tomb. He existed during these three days as a disembodied person.

Paul desires to be "absent from the body and to be at home with the Lord" (2 Cor 5:8), indicating that at death he will continue in God's presence apart from his body.[40] Likewise, he says,

> To live is Christ and to die is gain. But if I am to live on in the flesh, this will mean fruitful labor for me; and I do not know which to choose. But I am hard-pressed from both directions, having the desire to depart and be with Christ, for that is very much better; yet to remain on in the flesh is more necessary for your sakes. (Phil 1:21–24)

Paul clearly assumes that he can be absent from his body after death and yet continue to exist.

Furthermore, Paul assumes that he could continue to exist apart from his body, at least temporarily, *before* his death. In 2 Corinthians 12:2–4, he writes,

> I know a man in Christ, who fourteen years ago—whether in the body I do not know, or out of the body I do not know, God knows—such a man was caught up to the third heaven. And I know how such a man—whether in the body or apart from the body I do not know, God knows—was caught up into paradise.

Again, although the details of this passage are difficult to understand, it is at least clear that Paul, inspired by the Holy Spirit, believed it was possible that he could exist in a disembodied state—"out of the body"—but still be himself. In Cooper's words, "Everything points to his seriously entertaining the out-of-body option."[41]

---

observes that in Luke 24:35 Jesus "employed the word 'spirit' in distinction from flesh and bones in an attempt to reassure them: 'a ghost does not have flesh and bones, as you see I have' (v. 39). There is no doubt that Luke uses 'spirit' to mean 'disincarnate person.'" Cooper, *Body, Soul, and Life Everlasting*, 115. He discusses other passages that use *pneuma* on pages 112–15.

39. First Peter 3:18–19: "For Christ also suffered for sins once for all time, the just for the unjust, so that He might bring us to God, having been put to death in the flesh, but made alive in the spirit; in which He also went and made proclamation to the spirits in prison."

40. For a helpful commentary on this passage, see Witherington, *Conflict and Community in Corinth*, 391.

41. Cooper, *Body, Soul, and Life Everlasting*, 151.

The author of Hebrews speaks of the "great cloud of witnesses surrounding us" (Heb 12:1), referring to the saints who have died before us yet live on as "the spirits of the righteous made perfect" (Heb 12:23). According to Cooper, "There is no question here that the spirits are human spirits.... The dead exist apart from their earthly bodies."[42] Finally, Revelation 6:9–10 tells us that those martyred for their faith continue to live on, as disembodied souls who can nonetheless still reason, desire, and speak:

> I saw underneath the altar the souls of those who had been killed because of the word of God, and because of the testimony which they had maintained; and they cried out with a loud voice, saying, 'How long, O Lord, holy and true, will You refrain from judging and avenging our blood on those who live on the earth?'"

Taking all biblical passages into account, the Scriptures teach that we are a functional unity of soul and body but also an ontological duality, with the soul being the more fundamental aspect of what we are. Our souls are deeply united with our bodies in this present age. At some point our bodies will die, yet we will live on in a disembodied, intermediate state. However, this separation of body and soul is only temporary. At the end of the age, our bodies will be resurrected, our souls and bodies will be reunited, and we will live embodied, again as a functional unity as we were created to be, forevermore in the world to come.

CHRISTIANS THROUGHOUT THE AGES AGREE

As mentioned in the introduction, the vast majority of Christians over the past two millennia have shared this understanding of us as an ontological duality.[43] This helps to assure us that we are not reading into the text what we want it to say. Instead, we stand on the shoulders of the vast majority of believers who have studied the Scriptures and come to this same conclusion.

For instance, in the second century Mathetes observed, "The soul dwells in the body, yet is not of the body."[44] Tertullian concurs: "Certainly you value the soul as giving you your true greatness—that to which you

---

42. Cooper, *Body, Soul, and Life Everlasting*, 113–14.

43. For a thorough treatment of the history of this idea in Christian thought, see Goetz and Taliaferro, *A Brief History of the Soul*.

44. Mathetes, *Letter to Diognetus*, chapter 6 (1.27), available online at https://www.logoslibrary.org/mathetes/diognetus/06.html, accessed October 5, 2023.

belong; which is all things to you; without which you can neither live nor die."[45] Augustine writes, "All that does not live is without soul,"[46] and "I myself am a soul."[47] Aquinas adds, "There are myriad passages of sacred Scripture which proclaim the immortality of the soul."[48] Contemporary theologian Millard Erickson concludes that this view "was commonly held from the earliest period of Christian thought. Following the Council of Constantinople in 381, however, it grew in popularity to the point where it was virtually the universal belief of the church."[49]

Contrary to what some believe, this agreement cannot be attributed to the distorting influence of Greek philosophy on biblical interpretation and theology. Throughout church history Christians around the world who have had no exposure to Greek philosophy have interpreted the Scriptures as teaching that we are both soul and body. Given this nearly universal affirmation around the world and throughout the history of the church, we can have confidence in this truth of Scripture. Therefore, as Cooper sums up, "Christian philosophers and scientists need not adopt conceptual paradigms that implicitly contradict sound doctrine."[50] By this he is primarily referring to physicalism.

WHERE DID NEUROTHEOLOGY GO WRONG?

Neurotheologians seem to value integrating biblical truth with what we are learning from neuroscience. Yet the picture that emerges from a careful study of all the relevant texts of Scripture is deeply at odds with the physicalism of neurotheology. How did this occur? As we shall see, these neurotheologians made a fundamental philosophical assumption that is false, and dangerous, without any training to know they were doing so. In the next chapter, I will explore this wrong assumption.

---

45. Tertullian, *Soul's Testimony*, 6.
46. Augustine, *Immortality of the Soul*, 3.3.
47. Augustine, *Immortality of the Soul*, 30.61.
48. Aquinas, *Summa contra Gentiles*, II.79.17 Aquinas's nuanced view will be discussed in further detail in Chapter 6.
49. Erickson, *Christian Theology*, 540.
50. Cooper, *Body, Life, and Soul Everlasting*, xxvi.

3

Neurotheology's Wrong Assumption About Our Mental Life

"The first to state his case seems right until another comes and cross-examines him."
—Proverbs 18:17

"Knowledge is true representation of how things are."
—Dallas Willard[1]

> **CHAPTER SUMMARY**
>
> Assumptions determine how we interpret data, leading to the conclusions we draw and the applications we make. Therefore, having correct assumptions is of the utmost importance. A central assumption of neurotheologians, which they bring to the interpretation of neuroscience, is that when two things come together, they must be the same thing. Therefore, a neural event related to a desire, for instance, means that desire is nothing but that neural event. A number of reasons are given to show

---

1. Wilder, *Renovated*, 27.

> why this is a wrong assumption, drawing on philosophy to clarify relevant concepts and identify unsound reasoning. To show that mental events are not identical to neural events, I clarify the nature of identity, proving that we cannot be identical with our brains. From this philosophical data, I further confirm what the biblical data indicates: the physicalism of the neurotheologians is an inaccurate assessment of what we are.

JESUS' GREATEST COMMANDMENT IS to love God and love others. We love God through becoming Christlike—that is spiritual formation. And we love others by seeking to meet their needs—that is ministry. But how can we best be formed spiritually and minister to others? That depends on what we are.

Neurotheologians offer one answer, as discussed in chapter 1. Yet based on the biblical data discussed in chapter 2, we have good reason to believe that their conclusion about our essential nature is wrong. Therefore, their applications to spiritual formation and ministry are equally misguided.

Where did the neurotheologians go wrong? To answer this, we must identify their fundamental underlying assumptions, for any interpretation of data is only as good as one's starting assumptions. The interpretation, following from the assumption, in turn leads to a conclusion and specific point of application. Put simply: Assumptions → Interpretation of Data → Conclusion → Application.

If one's assumptions are correct, one is in a good position to accurately interpret the data, which in turn bodes well for producing correct conclusions and applications. However, if one's starting assumptions are wrong, the interpretation of the data, conclusions, and applications will almost certainly be wrong as well.

We constantly use this process to draw conclusions and make life applications, though we are usually not aware of each step in our thinking. For instance, some years ago I was almost arrested in Thailand due to a wrong assumption that ultimately led to a wrong application. I was on a bus in Bangkok and needed to pay the fare. I accidentally dropped a coin and saw it rolling away (the *data*). My *assumption* was that coins are just currency there, as they are in the United States. From this assumption and data, I *concluded* that I should step on the coin to stop it from rolling, in order to pay my fare. From this assumption, data, and conclusion came my *application*: I immediately slammed my foot down on the coin. At this moment, everyone on the bus gasped and looked at me in shock. Only then did I learn that coins in Thailand are *not* just currency. They bear the image of the king, who is highly honored. Therefore, to step on a coin is to symbolically step on the king, showing him great disrespect. This is a punishable offense. Had I started with this correct assumption, when I encountered the data of my coin rolling away my conclusion would have been much different—that I should treat it as I would treat the king. As a result, I would have waited for it to come to rest and only then carefully picked it up. (The authorities were gracious to me because I was a foreigner. They didn't arrest me, but they did give me a very stern warning not to step on a coin again!)

In the same way, the neurotheologians' interpretation of the data of neuroscience is only as good as their underlying assumption. It seems that neurotheologians begin with a fundamental, yet wrong assumption. By evaluating this assumption, we can discover the root of their wrong interpretation, conclusion, and application.

NEUROTHEOLOGY'S WRONG ASSUMPTION: WHAT IDENTITY IS

As mentioned in chapter 1, neuroscience has made some important discoveries. One discovery is that experiences such as having a thought, feeling an emotion, and making a choice (mental events) correlate with neural activity in various regions of our brain. This is data from the domain of neuroscience, at least for very simple mental states like feeling pain.[2]

But how should we *interpret* this data? What do these correlations *mean* about what we are and how we flourish? This is where things get tricky. Scientific data doesn't interpret itself. Interpreters interpret the

---

2. Neuroscience has not been able to provide evidence of correlation between more complex concepts, such as, for example, between the various understandings of what knowledge is in the modern era of philosophy and specific brain states. In fact, it is hard if not impossible to see how such correlations could be identified. See Bennett and Hacker, *Philosophical Foundations of Neuroscience*, 2:109–235.

data. And sometimes there are two or more interpretations that are equally consistent with the scientific data. For instance, when it was discovered that the universe is expanding,[3] some interpreted the data to mean that the universe is in an eternal series of expansions and contractions, of which the current expansion is just one. Others interpreted the data to mean that there is only one expansion: the universe began at a point in time and continues to expand. From the empirical data alone, both interpretations are equally plausible. They are *empirically* equivalent—both are based on the exact same observational data.

Neurotheologians have a fundamental assumption that guides their interpretation of the data. They assume that when two things happen together, they must be identical. In other words, they assume that *constant conjunction means identity*—that the two events are actually the same thing. Therefore, when a mental event and a brain event occur together, the mental event must be identical to the brain event.

This assumption of identity is evident throughout Thompson and Wilder's writings, as seen in passages of their books quoted in the introduction and chapter 1. Such a view is sometimes referred to as reductive physicalism (or simply physicalism), because each type of "mental" state (such as a memory) is fully reduced to, or is identical to, a corresponding type of brain state.[4]

Yet this is not the only explanation of this data of correspondence between mental and brain states. Actually, three different explanations can be offered, *all based upon and entirely consistent with the same data*. Moreland outlines these options related to feeling empathy:

> Consider, for example, the discovery that if one's mirror neurons are damaged, then one cannot feel empathy for another. How are we to explain this? Three empirically equivalent solutions come to mind: (1) strict physicalism (a feeling of empathy is identical to the firings of mirror neurons); (2) mere property dualism

---

3. This data is referred to as the "red shift"—the light we observe from all planets and stars is moving toward the red end of the light spectrum, indicating that each of these heavenly bodies is moving away from us. This implies that the universe is expanding outward in every direction.

4. This is known as the Type-Type Identity Thesis: a certain type of mental event (say, a memory-type event) corresponds to a certain type of brain event (a brain-type event). Other physicalists argue for the Token-Token Identity Thesis: Each individual "token" (instance) of that type (say, the memory of my son's graduation) always correlates to a specific brain instance (i.e., to a specific set of neurons firing). For more on this distinction, see Moreland and Craig, *Philosophical Foundations for a Christian Worldview*, 251–58.

(a feeling of empathy is an irreducible state of consciousness of the brain whose obtaining depends on the firing of mirror neurons); (3) substance dualism (a feeling of empathy is an irreducible state of consciousness in the soul whose obtaining depends (while embodied) on the firing of mirror neurons).[5]

The "identity" thesis is only one of these three possible explanations of the scientific data; the other two explain the data equally well.[6]

Based on their assumption that constant conjunction equals identity, Thompson and Wilder assume the identity thesis (the first option Moreland lists) is the correct explanation. Is this assumption justified? This cannot be determined by the scientific data, for the data supports the other two explanations equally well. Therefore, the assumption must be evaluated for what it is: a *philosophical* assumption about the nature of constant conjunction that is *brought to* the data discovered by science. And philosophical assumptions must be evaluated with the tools and methods of philosophy.[7] If these philosophical assumptions turn out to be correct, Wilder and Thompson's interpretation of the data follows, as does their conclusion and application.

One might object that evaluating this assumption is itself within the domain of science, and so no further philosophical evaluation is required. Two responses show why this objection fails. First, to claim that science should evaluate its own starting assumptions is to assume science is the only, or at least the ultimate, source of knowledge. This itself is a philosophical assumption known as scientism. This view, and its flaws, will be discussed in some length in chapter 7. But briefly, to make the philosophical claim that philosophy is irrelevant, and that science should be our only source of knowing how the brain and mind are related, is self-defeating.[8] Furthermore, if the determination of starting assump-

---

5. Moreland, "In Defense of Thomistic-Like Dualism," 106–7. For more see Moreland, "A Christian Perspective," 2–12.

6. These options will be discussed in more detail in later chapters.

7. Moreland summarizes, "[Concerning] how conscious states . . . are and are not in the body . . . the methods and findings of neuroscience are unable to address the question and, in general, are largely irrelevant to the central questions constituting philosophy of mind." Moreland, "Scientific Late Medieval Aristotelianism (Organicism)," 106. Identifying and evaluating the starting assumptions concerning the nature, aims, and methods of science, and therefore how to interpret the data of science, are more specifically questions in the philosophy of science. For a good overview of these issues, see Harre, *The Philosophies of Science*, 34–61.

8. A self-defeating claim is a logical fallacy in which one must assume something is true in order to show it is false, or assume it is false to prove it is true.

tions in science is determined by science itself, it becomes impossible to adjudicate between conflicting interpretations of data based on different starting assumptions, as is the case here. We need a second-order discipline that stands outside science to evaluate these starting assumptions. This is the role of philosophy.[9] It follows that we cannot discount the central role philosophy plays in evaluating Thompson and Wilder's starting assumption.

Therefore, are there good philosophical reasons to believe Thompson and Wilder's assumption is right? Neither of them offers us any justification for this assumption. They just begin with it as a given, and then they interpret the data accordingly. Neither Thompson or Wilder have professional training in philosophy, and so perhaps they are not aware of this assumption or its need for justification. In fact, this is a common failing among neuroscientists in general. As Dennis Noble, professor of physiology at Oxford University, writes in his foreword to *Philosophical Foundations of Neuroscience*, "Neuroscience has frequently and systematically confused conceptual and empirical questions."[10] The authors of *Philosophical Foundations of Neuroscience*, one a neuroscientist at the University of Sydney (M. R. Bennett) and the other a philosopher at Oxford University (P. M. S. Hacker), add in their introduction:

> Empirical questions about the nervous system are the province of neuroscience . . . conceptual questions (concerning, for example, the concepts of mind or memory, thought or imagination), the description of the logical relations between concepts (such as between the concepts of perception and sensation, or the concepts of consciousness and self-consciousness), and the examination of the structural relationships between distinct conceptual fields (such as between the psychological and neural, or the mental and the behavioural) are the proper province of philosophy. . . . Distinguishing conceptual questions from

---

9. For more on the nature of philosophy, see Moreland and Craig, *Philosophical Foundations for a Christian Worldview*, 12–14. They observe, "When philosophers examine another discipline to formulate a philosophy of that field, they ask normative questions about that discipline (e.g., questions about what one ought and ought not believe in that discipline and why), analyze and criticize the assumptions underlying it, clarify the concepts within it and integrate that discipline with other fields. . . . Philosophy is the only field of study that has no unquestioned assumptions within its own domain . . . for questions about the definition, justification and methodology of philosophy are themselves philosophical in nature. Philosophers keep the books on everyone, including themselves" (p. 13).

10. *Philosophical Foundations of Neuroscience*, xiii.

empirical ones is of the first importance ... any incoherence in the grasp of the relevant conceptual structure is likely to be manifest in incoherences in the interpretation of the results of experiments ... [and] much contemporary writing on the nature of consciousness and self-consciousness is bedeviled by conceptual difficulties.[11]

Given the need to distinguish the philosophical from the empirical and to provide adequate philosophical grounds to justify interpretations of the data of neuroscience, to philosophy we now turn.[12] In doing so, we will find that there are no good reasons to believe that the neurotheologians' starting assumption is correct. Rather, there are good reasons to believe just the opposite. Therefore, the conclusion and the application of neurotheology to spiritual formation and human flourishing do not follow.

MENTAL EVENTS ARE NOT IDENTICAL TO BRAIN EVENTS

To review, neuroscience offers us scientific data to be explained: correlations between brain states and mental states. The neurotheologians' philosophical assumption is that constant conjunction is the same as identity. Therefore, they conclude that the indications of a correlation between a brain state and a mental state mean that the two are actually the same thing. The mental state is reducible to the brain state. We are our brain.

To understand why Thompson and Wilder's assumption is wrong, we must first get a clear understanding of what identity is. Otherwise we

---

11. *Philosophical Foundations of Neuroscience*, 1–2, 7. Bennett is a first-rate neuroscientist, well qualified to offer this assessment. The biography on the book's back cover describes him as "Emeritus Professor of Neuroscience and University Chair at the University of Sydney, Founding Director of the Brain and Mind Research Institute, and Chair of the Mind and Neuroscience—Thompson Institute. He is the author and co-author of numerous books, including *The Idea of Consciousness, History of the Synapse, History of Cognitive Neuroscience*, and *Stress, Trauma and Synaptic Plasticity*. He is past President of the International Society for Autonomic Neuroscience and of the Australian Neuroscience Society." The first edition adds that he is "the recipient of numerous awards for his research in neuroscience, including the Neuroscience Medal, the Ramaciotti Medal and the Macfarlane Burnet Medal."

12. In fact, we have already started to wade into philosophical waters in the earlier sections of this chapter. In subsequent chapters, we will continue doing so as we evaluate the neurotheologians' conclusions and applications (chapter 4), the nature of the soul (chapter 5), the soul's relation to the body (chapter 6), objections neurotheologians may offer (chapters 7 and 8), and application of what we learn to how we flourish (chapters 9 and 10).

may easily confuse identity with something it is not and may become sidetracked.

There is a law of logic that defines the nature of identity. It states that two things are identical if, *and only if*, they share all their properties in common.[13] For instance, suppose your young daughter has a goldfish she dearly loves (she creatively named him Gil). One day, when she is at school, you notice Gil floating in the tank, dead as can be. Anticipating her anguish, you properly "bury" Gil and buy another goldfish that looks just like Gil (but is actually Gil2). When your daughter comes home from school, she rushes up to her room to say hi to Gil, as she does every day. But soon you hear her begin to sob. You immediately know that you were unsuccessful. Your daughter quickly identified properties of Gil2 that were not true of Gil, properties you did not notice. Perhaps Gil2 was a slightly different hue of gold. Or his mouth was shaped a bit differently. Or his movements were not quite the same. Whatever the difference, your daughter implicitly applied the Law of Identity, and because she discerned different properties, she rightly but sadly concluded this *could not* be Gil.

In the same way, we can discern whether a mental event is identical to a brain event by applying the Law of Identity. If the mental event and brain event share all the same properties, they are necessarily the same. But if one has even a single property that is different from the other, then they are necessarily *not* identical—they are necessarily *different* things.

Upon reflection, we find a number of properties of the mind that are not true of the brain. Due to space constraints, I'll discuss just three: our first-person perspective, our freedom of the will, and our rationality.

---

13. The nature of identity was described by Gottfried Leibniz (1646–1716) as the Law of the Indiscernibility of Identicals. This law states that necessarily, for any x and any y, if x is identical to y, then for any property P, P must be true of x and P must also be true of y: $\Box(x)(y)[(x = y) \to (P)(Px \equiv Py)]$. Therefore, if a property is not shared between two things, they are necessarily *not* identical. Of course, "identity" is often used in an informal, imprecise way, such as "This car is identical to the one I had in high school," which we know to really mean "This car is very similar to the one I had in high school." But neurotheologians are not saying brain states and mental states are "very similar" to one another. They are saying they are identical to one another. Therefore, we must understand the precise nature of identity, as expressed in the Law of Identity, to evaluate this claim.

Our First-Person Perspective

One property that is true of the mind but not the brain is our first-person perspective. I have direct, private, first-person access to my mental events. I know what I am thinking, feeling, or choosing in a way that is not available to anyone else. But I do not have private access to my brain events. Any number of neuroscientists can know better than me the facts about my brain activity. Therefore, due to the fact that mental events are "private" but brain events are "public," these different properties mean the mind cannot be identical to the brain.

To illustrate the difference between third-person and first-person perspectives, suppose that you are the world's leading audiologist. You know more than anyone about the process of hearing, including how the brain processes sounds. However, you are deaf. You have never personally heard a sound. Then, one day you wake up and hear birds singing! Do you now know more about hearing than you did the night before? You don't know anything more about the brain processes involved in hearing. But you now *also* know what it is like to hear. This first-person perspective is more than, and is not reducible to, the third-person perspective of hearing that you fully understood the night before.[14]

Related to this is the fact that you can't be wrong about what you are experiencing from a first-person perspective, though others can be wrong from their third-person perspective.[15] Thompson gives an example when discussing the reality of emotion (though he does not apply the Law of Identity so as to properly conclude that the mind and brain must be different entities). He observes, "Emotion is not debatable. If your daughter senses the feeling of joy, shame, disappointment, or some general form of distress, that is in fact what she feels."[16]

Note that this is true even if she experiences this emotion while having brain surgery. Though the surgeon knows more about what is going on in that region of her brain than she does, each time he touches that region with an electrode he must ask for her first-person report: "What are you feeling now?" He believes he knows what she is feeling due to his

---

14. This is sometimes referred to as the knowledge argument. See Taliaferro, "Substance Dualism: A Defense," 49–53. For a discussion of specifically why Christian physicalism fails to explain first-person subjectivity, see the article in the same volume by Menuge, "Why Reject Christian Physicalism?" 395–400.

15. This is referred to as the incorrigibility of one's own mental states.

16. Thompson, *Anatomy of the Soul*, 95. See page 105 for another example.

stimulation of that region and the correlation he believes to exist between it and an emotion. But he may be wrong. On the other hand, she cannot be wrong about what she is feeling. So he asks her to verify what he thinks is the case, but *only she* knows for sure.

Finally, for every mental state, such as pain or the taste of a lemon, there is a what-it-is-like to that state that defines its nature. This first-person feeling of a pain is essential to what it is. But no physical state, including a brain state, has a what-it-is-like that characterizes it.[17]

With all this in mind, recall the Law of Identity: Necessarily, if two things are identical, they will *always* share *all* properties in common. Therefore, all that is required to show two things are not identical is one example of a property of the mind that is not also a property of the brain. First-person subjectivity is one such property, as I have just illustrated in several ways. Therefore, we can rightly conclude that the mind cannot be identical to the brain. As T. L. S. Sprigge observes, "The main reason for holding [that there is a distinction between the mental and physical] is that it seems entirely possible that a scientist should have complete knowledge of a human organism and a physical system and yet be ignorant of the special character of that individual's consciousness."[18]

---

17. For more on the direct knowledge of mental states such as these, see Hopp, *Phenomenology: A Contemporary Introduction*, which includes a helpful synopsis of Dallas Willard's thoughts on this fact. For more see Geniusas, *The Phenomenology of Pain*, and Rickabaugh and Moreland, *Substance of Consciousness*, 103–4.

18. Sprigge, *The Importance of Subjectivity*, 9. For more, see Moreland and Craig, *Philosophical Foundations for a Christian Worldview*, 234–39; Menuge, "Why Reject Christian Physicalism?" 395–400.

Bennett and Hacker, in *Philosophical Foundations of Neuroscience*, call the assumption that a first-person experience can be reduced to a brain event the Mereological Fallacy: "Mereology is the logic of part/whole relations. The neuroscientists' mistake of ascribing to the constituent part of an animal attributes that logically apply only to the whole animal.... Human beings, but not their brains, can be said to be thoughtful or thoughtless ... to see, hear, smell, and taste things; people, but not their brains, can be said to make decisions or to be indecisive. [These] have no intelligible application to the brain" (p. 73). In footnote 13 on page 73, the authors observe, "To be sure, this mereological confusion is common among psychologists as well as neuroscientists." I would add that it is common among neurotheologians as well. Note this fallacy also occurs (during our embodied state) when activities such as thinking are reduced to the soul alone, denying the important role of the brain due to our deep functional unity. The intricate relationship between the mind and brain in activities such as thinking will be discussed in chapter 6. (As discussed in chapter 2, since we will continue to exist as disembodied souls capable of thinking during the intermediate state, this fallacy is not applicable during that period.)

Although we need to find only one property not shared by both the mind and the brain to prove that they are not identical, I'll offer two more to further make the point.

Our Free Will

We Make Choices

A second thing true of us is that we make choices. We know that we have the capacity to select between various options independent of any prior conditions that are out of our control.[19]

The fact that we make choices seems to be known by everyone. Thompson and Wilder indicate that they know this to be true. For instance, Wilder's *Renovated* is based on his journey to learn from Dallas Willard. He states, "Dallas was proposing a practical shift in theology.... I concluded that my relationship to God needed more attachment love."[20] The rest of his book discusses the subsequent choices he made in order to learn from Dallas and develop greater attachment love. Furthermore, the fact that he wrote this book indicates a belief that others who read it can also choose to learn from Dallas and change their lives.

Thompson also assumes that we are free to make choices. He begins *Anatomy* with an invitation: "This book invites you to trust while reading the text."[21] He assumes that the reader can choose either to trust or not trust. The rest of the book is based on this choice to trust him, as he suggests ways in which the reader can make other choices, based on what he writes, to grow in Christ.

Free will is also assumed in the second main finding of neurotheology mentioned in chapter 1: neuroplasticity. As Thompson defines this, it is the brain's "capacity, at a cellular connection level, to make new synapses and to prune away those synapses that don't get much firing

---

19. I am assuming what seems to be the "common sense" idea of freedom: that I am free only if I could have chosen to do otherwise. If someone is holding a gun to my head or otherwise compelling me to "choose" something, it is not really a choice. This is known as libertarian free will, or counter-causal freedom. For a fuller defense of this as the most appropriate way to understand freedom, see Moreland and Craig, *Philosophical Foundations for a Christian Worldview*, 267–84; Rickabaugh and Moreland, *Substance of Consciousness*, 234–71.

20. Wilder, *Renovated*, 6–7.

21. Thompson, *Anatomy of the Soul*, 9.

action."[22] He then frames the question he will answer concerning this "pruning" of our brains: "What happens when we begin to consider that we can change the way our brains are wired?"[23] Free will is central to this thesis: we can choose (or not choose) to reshape our brains. Thompson invites us to make such a choice repeatedly: "I invite you to practice this meditation . . . [in which] you are . . . changing the neural networks of your brain,"[24] "Our neurons can be redirected in ways that correlate with joy, peace, patience, kindness, goodness, faithfulness, gentleness, and self-control,"[25] and again, "This neuroplasticity can be enhanced and facilitated by our intentional behavior."[26] Choosing to reshape our brains presupposes free will.

Finally, the fact that some argue the mind is identical to the brain in itself assumes free will. Those making this argument assume that they have weighed the evidence and chosen to believe what it supports. Furthermore, they believe those who read their books can also choose to believe as they do. This makes sense only if we do indeed have a choice in the matter—if we have freedom of the will.

Free will is undeniable. We assume it. Neurotheologians assume it. All who argue that the mind and brain are identical assume it. The question is whether our freedom to choose is a property of the mind *and* the brain (indicating identity), or a property of the mind *but not* the brain (indicating difference).

Our Brains Cannot Make Choices

Brains are physical things. Therefore, like all other physical things, the brain's operations are determined by the laws of physics and chemistry: when X conditions are present, Y happens. There is no other option. For instance, consider physical things like the balls on a pool table. When I hit the cue ball with a certain velocity and from a certain direction, and when it contacts the 8-ball, the trajectory of the 8-ball is determined by those prior factors. It has no choice in the matter. Notice that it makes no difference how big or small the physical elements are, how fast they

---

22. Thompson, *Anatomy of the Soul*, 45.
23. Thompson, *Anatomy of the Soul*, 48.
24. Thompson, *Anatomy of the Soul*, 143.
25. Thompson, *Anatomy of the Soul*, 87.
26. Thompson, *Anatomy of the Soul*, 46, quoting Daniel Siegel (no citation provided).

move, or how complex their configuration is. Regardless, the laws of physics control the outcome. This is what makes playing pool and doing science possible. As David Papineau says,

> All physical events are determined, or have their chances determined, by prior *physical* events according to *physical* laws. In other words, we never need to look beyond the realm of the physical in order to identify a set of antecedents which fixes the chances of subsequent physical occurrence. A purely physical specification, plus physical laws, will always suffice to tell us what is physically going to happen.[27]

Now let's compare this to the realm of neuroscience. When atoms collide in brains, the results are equally determined by the laws of physics and chemistry. Concerning free will in particular, William Hasker states that if physicalism is true, "What we do is simply the result of what those particles do—there is nothing there but the particles to 'do' anything. It is clear, then, we have no real choice about what we do; even if we have the 'experience of choosing,' how that choice comes out is wholly determined by the actions and reaction of the fundamental particles."[28]

Yet as discussed above, our choices are *not* determined. That's what makes them choices. Therefore, free will cannot be a property of the

---

27. Papineau, *Philosophical Naturalism*, 16. Some may object that there are fluctuations at the quantum level that explain freedom physically. However, this alternative is untenable. First, quantum physics is a relatively new field, and so we should not assume that because we do not yet *know* the causes of quantum events, there are in fact no causes. It is more reasonable to conclude that, since the material world operates by cause and effect at the macro level, the quantum level does as well, and continue our scientific investigation to discover these yet unknown causal factors. There may turn out to be metaphysical indeterminacy at the quantum level. But at this point it is simply too early to reach such a conclusion. Yet suppose quantum indeterminacy *is* a metaphysical reality (i.e., it is how things really are) rather than an epistemological reality (what we currently know of this relatively new field of study). In this case, quantum fluctuations are not caused. They are entirely random. But as the one raising this objection assumes, these quantum laws of physics, while themselves not determined, do in turn determine the resulting macro-effects in the brain. So human actions, if this is right, are still determined. Therefore, this objection fails to provide a physical explanation of free will. See Moreland and Craig, *Philosophical Foundations for a Christian Worldview*, 278–79; Swinburne, *The Evolution of the Soul*, 235, 238–47.

28. Hasker, "On Behalf of Emergent Dualism," 84. Note that moral choices are one type of choice—choosing to do what is right or wrong. Therefore, if our choices are identical with neural events, virtue also loses all meaning, as does sin. So in this same passage, Hasker rightly asks, "If we have no free will, how can we be responsible? In particular, how can we be responsible before God, as Scripture says we are?"

brain. It must be a property of the mind. So the mind and brain cannot be identical.[29]

Our Rationality

We Reason

Finally, we are also rational beings. Reason can be defined as the ability to understand and evaluate arguments, as well as to avoid logical fallacies.[30] Granted, sometimes we do better at reasoning than at other times. But the fact that we or others can see where our reasoning goes wrong, and that we are able to correct our reasoning, is further evidence that we possess the property of rationality.

Thompson and Wilder certainly assume we have rationality. The very fact that they bothered to invest the time and energy in writing their books containing arguments for neurotheology proves this. They assume that they have evaluated the data and come to the most reasonable conclusion concerning, as Thompson puts it, the "surprising connections between neuroscience and spiritual practices that can transform your life and relationships."[31] Similarly, Wilder has rationally concluded how we can be *Renovated* and is seeking to help us draw logical connections between, in the words of his subtitle, *"God, Dallas Willard and the Church That Transforms."*

---

29. See Moreland and Craig, *Philosophical Foundations for a Christian Worldview*, 240–43; Moreland and Rae, *Body and Soul*, chapter 3, "Human Persons in Naturalist and Complementarian Perspectives," 87–120 and chapter 4, "Substance Dualism and the Human Person: Free Agency," 121–56; Goetz, "Naturalism and Libertarian Agency," in Craig and Moreland, eds., *Naturalism: A Critical Analysis*, 156–86; Rickabaugh and Moreland, *The Substance of Consciousness*, 144–88. Willard, "Knowledge and Naturalism" in the same text, 24–48; Willard, "Knowledge," in Smith and Smith, eds., *The Cambridge Companion to Husserl*, Smith and Smith, 138–67.

30. An argument is simply a series of statements (premises) that lead logically to a conclusion. Reasoning involves (1) understanding what these premises claim, (2) evaluating whether they are accurate claims about reality, and (3) determining whether, if the premises are true, the conclusion logically follows. If so, it is reasonable to act on that conclusion, for it is reasonable to believe it is true (it corresponds to reality). A logical fallacy is an instance of faulty logic, such as begging the question (i.e., building your conclusion into your premises so that you inevitably get the conclusion you want).

31. From the subtitle of *Anatomy*.

Our Brains Do Not Reason

But similar to our argument above about free will, brains also don't have what it takes to reason. Brain function is determined by the laws of physics and chemistry. Everything said earlier concerning free will is also true of rationality. It makes no sense to say that the 8-ball deduced rationally that it should go in a certain direction at a certain velocity. That is not what material things do.

Therefore, given our ability to reason, and the fact that material things cannot reason, mental states clearly have a property that differs from brain states. From the fact of rationality it follows, by the Law of Identity, that minds and brains are necessarily different.[32]

We have considered three types of mental states true of all human beings: first-person subjectivity, freedom of the will, and rationality. We have seen that all three differ from brain states.[33] Therefore, given the Law of Identity, Wilder and Thompson's fundamental assumption that constant conjunction means identity, and therefore their interpretation of neuroscientific data to mean mental states are identical to brain states, is false.

ARE NEUROTHEOLOGIANS A NONREDUCTIVE TYPE OF PHYSICALIST?

Yet there is a form of physicalism that doesn't identify mental events with brain events. It is known as *nonreductive physicalism*.[34] According to this view, the brain is still the ultimate reality. But brains, due to their complexity, can secrete mental events (such as choices) as by-products. These

---

32. See Groothuis, *Christian Apologetics*, 398–99; Lewis, *Miracles*, 16–28; Moreland, *The Recalcitrant Imago Dei*, 67–103.

33. There seem to be a number of other, similar mental events that are not identical to brain events. For instance, finding guidance in dreams by God, an experience recorded repeatedly in the Scriptures, loses all meaning if identical to the brain. The same is true of the creativity of a painter or the imagination of an author.

34. See Murphy, "Nonreductive Physicalism," 115–38. Nonreductive physicalism is currently the most popular form of property dualism, which holds there is one *substance*—one underlying thing—which is the brain. Yet this one substance has two types of properties: physical properties, such as weight, and mental properties, such as choices. Similarly, it may be referred to as *epiphenomenalism*, according to which mental properties such as choices are epiphenomena: they exist as separate from the brain but are caused by and are dependent on the brain for their existence (they "supervene" on the brain). This term calls out the causal connection: the material reality (the brain) is responsible for the production (the emergence) of the immaterial reality (the choice).

mental events depend on the brain for their existence, but they are not identical to brain states. To state the position more technically, there is not ontological *identity* (neural events and mental events are truly different things), yet there is ontological *dependency* (mental events depend on brain events for their being).

An analogy is the relationship between smoke and fire. Smoke and fire are different things. We know this due to the Law of Identity: they have different properties, such as different temperatures. Therefore, they cannot be identical. However, smoke is produced *by* fire. It is a by-product of fire, which is the underlying, ultimate reality. We see this when the fire goes out. No fire, no smoke.

That's why this view is termed *nonreductive* physicalism. It is a variety of physicalism, in that ultimate reality is physical—the brain. Yet it is nonreductive, in that it doesn't treat mental events as fully reducible to brain events. Mental events *are* different. Yet they can exist only if produced by and sustained by a brain. No brain events, no mental events. Simply put, whereas reductive physicalism says a thought is identical to (i.e., nothing but) neurons firing, nonreductive physicalism says a thought is different from, though still entirely caused by, neurons. Yet both are physicalist understandings, for in both cases the thought is ultimately due to the operation of the brain, not the mind.

Is it possible that this is Wilder and Thompson's view, rather than reductive physicalism? Perhaps. On one hand, they regularly make statements equating the mind with the brain, as discussed in the introduction and chapter 1. Yet on the other hand, several passages in *Renovated* and *Anatomy* are ambiguous. When we are determining their anthropology, it seems appropriate to interpret these few and more ambiguous passages in light of the far greater number of passages that clearly state the identity thesis of reductive physicalism. But for sake of argument, let's consider whether Thompson and Wilder may actually be nonreductive physicalists rather than reductive physicalists, and the implications if this is the case.

In one passage, Wilder says, "The human-identity systems in the brain *generate* our emotional reactions to life."[35] Perhaps by "generate" he means that our emotional reactions emerge from and are distinct from our brains. Elsewhere he says, "Mind comes from the same brain structures that produce mindfulness and mindsight."[36] The terms "comes from" and

---

35. Wilder, *Renovated*, 3 (emphasis added).
36. Wilder, *Renovated*, 37.

"produce" may further indicate a belief that these are by-products of the brain, different not in degree but in kind from the brain event. Finally, in one footnote Wilder does distance himself from reductive physicalism.[37] As mentioned in chapter 1, Wilder did study at Fuller Seminary where Nancey Murphy and Joel Green, both leading nonreductive physicalists, teach. Perhaps they influenced him to embrace nonreductive physicalism as well, though he is confusing the two forms of physicalism in his book, often speaking as a reductive physicalist.

Thompson may also be a nonreductive physicalist. At one point in *Anatomy of the Soul*, he states, "The brain/mind matrix is considered by some researchers to be a primary example of complex systems among living things . . . small interconnected parts (neurons) interact to form larger parts (neural networks), which are completely different from the neurons themselves."[38] He is not clear as to whether these neural networks are "completely different" in terms of their material constitution or complexity (a difference in *degree*) or in terms of having emergent properties (a difference in *kind*). If it is the latter, he is speaking here as a nonreductive physicalist.[39] Elsewhere he states, "The limbic circuitry . . . is the wellspring of primal neural activity that eventually *emerges* . . . in the form of fear, joy, disgust, anger, hurt, disappointment, relief, and dozens of other emotions."[40] Here again, it is unclear what he means by "emerges," but it may be an endorsement of nonreductive physicalism. When discussing shame, he claims that it "is a function of the autonomic nervous system's balance between sympathetic and parasympathetic fibers. . . . It *emerges*

---

37. Wilder, *Renovated*, 216, footnote 19: "Neither Dallas nor I would accept that the cingulate is the soul. The cingulate is in the body. We would both believe that the body should harmonize with the soul." This is certainly true of Dallas Willard, who believed that the soul is ontologically distinct from the body. But in light of Wilder's many other reductionist passages in *Renovated*, it is not clear what he means to say here.

38. Thompson, *Anatomy of the Soul*, 237.

39. Yet in summarizing this section he goes on to say, "This concept leads us to the pinnacle theme of this book: when our brains operate in a flexible, adaptive, coherent, energized, and stable fashion, we are able to live in *community* in a way that encourages those around us to develop these same qualities." Thompson, *Anatomy of the Soul*, 238. This seems to indicate that he sees all he is writing about in this section, including complexity, as features of brains, and so in the earlier passage in question he is referring only to a difference in degree. If so, this passage is not an example of nonreductive physicalism. It is hard to know precisely what he does and does not intend to affirm here.

40. Thompson, *Anatomy of the Soul*, 39 (emphasis added). The limbic area of the brain is a set of brain structures surrounding the boundary between the cerebral hemispheres and the brainstem.

in the presence of a dis-integrated prefrontal cortex."[41] Again, perhaps he is using "emerges" to indicate nonreductive physicalism. Finally, at one point he states, "I am in no way attempting to explain those ideas [about following Jesus] on the basis of the brain; I am not reducing them to the function of neural networks."[42] Similarly, he claims that we can't use neuroscience to "reduce" our desire for God to neurons, contrary to what atheist scholars such as Steven Pinker and Daniel Dennett are saying.[43] Given the many places throughout *Anatomy* where he does in fact clearly reduce mental events to neural events,[44] these claims seem confused. But perhaps he genuinely does not intend to be reductionistic in his writings.

Therefore, for the sake of argument and based on these few passages, let's suppose Wilder and Thompson are in fact nonreductive physicalists. Does this provide a helpful way forward for them? It does not. Let's see why.

INADEQUACIES OF NONREDUCTIVE PHYSICALISM

Freedom Is Still Excluded

Let us again consider freedom of the will. The nonreductive physicalist does not claim that a choice is identical to a neural event. The choice has different properties, such as being immaterial. Yet, while not identical to the neural event, the choice is still dependent on and determined by the underlying neural event that caused it. Again, fire generates smoke as a

---

41. Thompson, *Anatomy of the Soul*, 212 (emphasis added).

42. Thompson, *Anatomy of the Soul*, 185.

43. "Several prominent scholars, such as Steven Pinker and Daniel Dennett, have in fact attempted to use [neuroscience] to *disprove* the reality of God and the validity of religious experience. It seems to me that one way to express their perspective is to say that if we can reduce our experience (in this case, of God) to that which we can measure (our genes and our neurons), we can eliminate the necessity of the God we thought existed." Thompson, *Anatomy of the Soul*, 10.

44. In the passage just cited Thompson immediately turns around and does just what he criticizes Pinker and Dennett for doing: reducing beliefs to brain states. He claims, "Most of us either want to believe in and have a relationship with God or we don't. Either way, we'll find ways for our left hemispheres to 'prove' what our right hemispheres are longing for—or are too terrified to desire." Thompson, *Anatomy of the Soul*, 10. His point is that the left brain's rationalism must not override the right brain's intuitions and "big picture" thinking about God (which is what he is implying that Dennett and Pinker are doing). Yet, either way, Thompson is reducing beliefs to functions of the brain (one hemisphere or the other), contrary to his immediately prior claim that Dennett and Pinker should not do so. It is hard to know what to make of this.

by-product. So too, on this view, neural events produce mental events, such as choices, as by-products.

But how can something material create something immaterial? One key principle of logic is that "out of nothing, nothing comes." In other words, a cause must first possess that which it passes on. For instance, a person must first have at least $500 in his bank account before he can write a check for $500. Anything less is insufficient, and the check will bounce.

The brain is purely material. Therefore, it possesses only the ability to create other material things—it can only "pass on" physical properties, because that is all it has to give. This is not just a logical deduction, but it is also apparent from scientific study. In our investigations of the universe, we never find material things giving rise to immaterial things. Hasker notes that if we want to believe mental properties can emerge from the brain, "we must then ascribe to the physical substance properties quite unlike those it is known to have *in all other contexts*."[45]

What reasons might be given to support this idea that just this once—in the case of brains—the material creates the immaterial? Thompson or Wilder offer none. It seems that there are no reasons. As the physicalist Jaegwon Kim says, if we want to believe that this occurs, it "will forever remain a mystery; we have no choice but to accept it as an unexplainable brute fact."[46] Yet unless one is committed to physicalism in order to justify other commitments (such as atheism), just accepting this on blind faith seems too high a price to pay.

For the sake of argument, let's suppose that brains *can* produce immaterial things such as choices. This leads to an even bigger problem (if that is possible!). In this case, the choice is by definition a by-product of the brain. But by-products are not the type of things that can turn around and cause their source to do anything. Smoke cannot cause a fire to burn more brightly, or go out. The causal chain runs only one way: from the

---

45. Hasker, *Metaphysics*, 71.

46. Kim, *Philosophy of Mind*, 229. On this tendency among physicalists to confuse ignorance with mystery, Bennett and Hacker observe that it is often based on confusions generated by false philosophical assumptions (what they call "confused ideas"). "Ignorance is one thing, mystery another. . . . One should be wary when told that something is a deep mystery. There are many subjects about which scientists are ignorant, and many empirical questions to which they do not know the answers. . . . One may be too hasty in declaring something to be a 'mystery.' For in some cases, we do not merely have no clear idea how to discover the truth about a certain subject, we have thoroughly confused ideas." Bennett and Hacker, *Philosophical Foundations of Neuroscience*, 241–42. By "confused ideas," they are referring to assumptions that are not sufficiently justified philosophically.

fire to the smoke. Similarly, if a "choice" is somehow produced as a by-product of the brain, it wouldn't be able to turn around and act on the brain (say, by causing neurons to fire that lead to raising your hand to vote). The causal chain can run in only one way: from the brain event to the mental event.[47]

Therefore, even if there is some rational way to defend the idea that the brain can cause mental properties such as choices, they turn out to be inert. They don't help the nonreductive physicalist explain the reality of our choices in action.[48]

Rationality Is Still Excluded

The nonreductive physicalist further posits that once the brain develops a certain level of complexity, mental properties of rationality also emerge. Again, these mental properties are different from the properties of the brain but are fully dependent on the brain for their existence and sustenance.

Nonreductive physicalism faces similar problems in making sense of rationality as it did in making sense of freedom of the will. First, it is no more plausible that rationality can emerge from the physical properties of a brain state than freedom can emerge. Yet even if we again suppose this could somehow happen, such a by-product of the brain state isn't the type of thing that can turn around and cause further brain events (such as causing neural events that would in turn produce a subsequent action I believe to be "reasonable"). Reason, as a by-product of the brain, would also be inert.

Furthermore, let's assume that somehow reasons emerge from complex neural events. And then let's further assume that these reasons can in turn cause brain events. Even if so, a third problem emerges: if this is the case, we would have no basis to believe that our reasoning should be relied upon to draw true conclusions and guide action.

We trust our conclusions and act on them because we are convinced that our process in coming to these conclusions was rational—that our

---

47. Believing that epiphenomena can in turn have causal influence on the brain is known as downward causation. For a further discussion of this view, and more detailed critiques, see Hasker, *In Search of the Soul*, 87–89, including his footnote 18 which cites his debates with Murphy on this topic in *Philosophia Christi* 2, no. 2 (2000). See also Rickabaugh and Moreland, *Substance of Consciousness*, 177–78.

48. The sources cited above in reference to reductive physicalism's inability to explain freedom also address the inadequacies of nonreductive physicalism to explain rationality (see below).

response to the truth of the premises was the right one, logically speaking. This is apparent in the language we use, such as "I examined the evidence and concluded that such-and-such is the case" or "I realized that the data supported this conclusion."

However, nonreductive physicalists believe the ultimate cause of all mental events, including events of reasoning, is the molecular activities of our brains. Yet the activities of molecules are, by definition, nonrational. Therefore, we cannot expect that any by-product that results from nonrational matter would be rational. "Out of nothing, nothing comes." If we are drawing conclusions only in virtue of these material causes, governed by the laws of physics and chemistry, we should not trust them to be true conclusions.

This is not to say that, if reason is ultimately produced by the brain, it may not *happen* to give us true conclusions now and then. However, that result would be purely coincidental. We would have no grounds for basing any of our beliefs or actions on the deliverances of reason, if this were the case. As J. B. S. Haldane admits, "If my mental processes are determined wholly by the motions of atoms in my brain, I have no reason to suppose that my beliefs are true. They may be sound chemically, but that does not make them sound logically."[49]

In sum, even if neurons could produce "reason," and if this "reason" could turn around and cause subsequent neural events, we would have no warrant to believe these "reasons" were rational, reflecting truth. Therefore, we should not, in turn, use these reasons as the basis for other decisions and actions (such as for accepting nonreductive physicalism).[50] But this is the whole point of rationality. Nonreductive physicalism again is no better than reductive physicalism.[51]

---

49. Haldane, "When I Am Dead," 209.

50. See Menuge, "Why Reject Christian Physicalism?" 394–410; Groothuis, *Christian Apologetics*, 400–408; Lewis, *Miracles*, 16–28. Some may suppose that evolution selects for truth-finding abilities. However, evolutionary theory holds that selection is based on survival value alone. Therefore, whether the belief is true is irrelevant, as long as it produces survival (such as a person believing that rattlesnakes are harmless, yet also believing that he should never pick up what is harmless). As physicalist Steven Pinker puts it, "Our brains were shaped for fitness, not for truth." Pinker, *How the Mind Works*, 21. See also Alvin Plantinga's "Evolutionary Argument Against Naturalism," versions of which appear in a number of places in print and online. It is outlined in some detail in his *Warranted Christian Belief*, 227–40. He and others discuss the argument further in Beilby, *Naturalism Defeated? Essays on Plantinga's Evolutionary Argument Against Naturalism*.

51. Similar arguments can be offered against physicalism from the mental properties

In sum, Thompson and Wilder assume that due to the constant conjunction between neural events and mental events, the two are identical (reductive physicalism). This assumption has been shown to be false. Yet they are a bit ambiguous in their writings as to the exact nature of mental events, leaving open the possibility that they are nonreductive physicalists. However, this alternative assumption is no more helpful in making sense of mental states such as freedom and rationality. In the next chapter, I'll draw out the wrong conclusions and applications that follow from their physicalist assumption, regardless of whether reductive or nonreductive physicalism is assumed.

---

of sensations, propositional attitudes (attitudes toward extramental states of affairs), intentionality (that our thoughts are always "of" or "about" something), beliefs, feelings, and so on. See DeWeese and Moreland, *Philosophy Made Slightly Less Difficult*, 105–29. For a more detailed treatment, see Moreland and Craig, *Philosophical Foundations for a Christian Worldview*, 228–46; See also Swinburne, *Evolution of the Soul*, 17–121. For a detailed discussion of the issues in this chapter and other reasons to reject a physicalist anthropology, see Rickabaugh and Moreland, *The Substance of Consciousness*.

4

Neurotheology's Wrong Conclusion about What We Are

Reality can be described as what we humans run into when we are wrong, a collision in which we always lose.
—Dallas Willard[1]

Progress means getting nearer to the place you want to be. And if you have taken a wrong turn, then to go forward does not get you any nearer. If you are on the wrong road, progress means doing an about-turn and walking back to the right road.
—C. S. Lewis[2]

> **CHAPTER SUMMARY**
>
> From the neurotheologians' wrong assumption identified in chapter 3, this chapter evaluates the wrong conclusion they draw: that we are ultimately a body, rather than a soul. This conclusion is shown to be wrong for two reasons. First, our many experiences of the world are

---

1. Willard, *The Allure of Gentleness*, 3.
2. Lewis, *Mere Christianity*, 36.

> bound together into a unified whole. This unity cannot be the result of our brain. It can only be the work of our soul. Second, each of us has a past, present, and future. Yet the brain, like all parts of our body, is changing at every moment and is completely replaced roughly every seven years. Therefore, the body/brain cannot account for the continuity of all our past and our future experiences. Only if we are a soul can we make sense of our being the same person from birth to death and into the afterlife. From the neurotheologians' wrong conclusion follows their misguided application to human flourishing. *Neural* formation is not the answer. Rather, *spiritual* formation is the answer.

THE NEUROTHEOLOGIAN'S CONCLUSION

IN CHAPTER 3, WE explored Curt Thompson and Jim Wilder's idea that when we speak of mental events, we are really talking about brain events. My brain, they say, is the "haver" of all my experiences. This leads to their inevitable conclusion: Since it is *I* who have my experiences, and they believe *my brain* has these experiences, they conclude that "*I*" am *my brain*.

Ascribing to the brain the role of integrating all our parts into one person, creating our identity, is repeated often in *Renovated* and *Anatomy*. As noted earlier, Wilder's writes, "A supraconscious brain process stays ahead of conscious thought . . . the brain is constantly calculating the answer to '*Who am I now*?'"[3] and "Our brain creates and maintains a human identity."[4] He adds that "Dallas [Willard] describes the soul as 'that part of the person that integrates all the other dimensions to make one life.' The brain happens to contain a structure whose function is the integration of all internal states and external connections with others. . . . When Dallas describes our experience of the soul . . . he could hardly have described the cingulate [cortex] in clearer terms."[5]

3. Wilder, *Renovated*, 36.

4. Wilder, *Renovated*, 68.

5. Wilder, *Renovated*, 85. See chapter 7 for a discussion of Wilder's misinterpretation of Willard's anthropology.

Again, Thompson echoes this thinking. He writes it is the prefrontal cortex "that sets us apart from all of God's other created beings. Attention, memory, emotion, and attachment all come together and are integrated at the PFC"[6] and again, "The left hemisphere sets me apart as 'me.'"[7] He adds, "The heart—our deepest emotional/cognitive/conscious/unconscious self—is manifest most profoundly at the level of the prefrontal cortex,"[8] and "the reptilian, limbic, and cortical portions of our brains [are] those parts of our souls by which God's voice is mediated."[9] Elsewhere he states, "When we think about what 'makes us human' biologically . . . we think of the PFC [prefrontal cortex]. . . . Not only can we learn a lot about what makes us human by studying the anatomy of the brain, we acquire additional insights from considering its various *systems.*"[10] For Thompson and Wilder, our brains are what ultimately make us what we are.

Furthermore, Wilder's main point in *Renovated* is grounded on this conclusion. His thesis is that we most need "attachment love"—healthy, loving relationships with others, including God. And since we are our brains, the brain needs to be attached in such loving relationships. For Wilder, it is "the brain system that forms and changes character . . . through attachment bonds."[11]

Thompson agrees. He states, "I will argue that it is only through this process of being known that you come to know yourself and learn how to know others . . . taking in all the nonverbal cues that lead, in secure attachments, to a fully functioning prefrontal cortex."[12] In effect, the prefrontal cortex is me, and therefore it must be fully functioning for me to be known and know others in secure and healthy relationships.

---

6. Thompson, *Anatomy of the Soul*, 157. Historically, our uniqueness has been understood to be the result of our bearing the image of God. I will discuss this point in detail in chapter 5.

7. Thompson, *Anatomy of the Soul*, 244.

8. Thompson, *Anatomy of the Soul*, 169.

9. Thompson, *Anatomy of the Soul*, 205. The triune brain model divides the brain into these three parts.

10. Thompson, *Anatomy of the Soul*, 40, 41. Given his physicalism, it is unclear why he adds "biologically" here, as if there were another way to speak of a person from his perspective. This is one of a few places in *Anatomy* where he is ambiguous and possibly inconsistent in articulating his anthropology, and perhaps nodding in the direction of nonreductive physicalism. This chapter will also explore why nonreductive physicalism is an inadequate explanation of what we are.

11. Wilder, *Renovated*, 39, 124.

12. Thompson, *Anatomy of the Soul*, 23, 247.

TWO REASONS WHY THIS CONCLUSION CAN'T BE RIGHT

If I am a brain, in contrast to the historic view that I am an immaterial soul that has a brain, two implications follow. This chapter will discuss these implications in detail and show why these implications should lead us (in addition to the biblical data discussed in chapter 2) to reject the neurotheologians' conclusion that we are our brains.

Brains Cannot Explain Our Unity at Each Moment

At any given time, we know we are a unity of many experiences. Yet, as I argue in what follows, if we are a brain, we cannot be such a unity at any given time. Only a soul can unify these experiences. Therefore, based on the data of our everyday experiences, we must not be a brain, but a soul.[13]

I Have Unified Experiences

Concentrate on an object that is near you. What do you experience? Right now, I am focusing my attention on an item on my desk. I see it is blue, cylindrical, about four inches tall and three inches in diameter. As I pick it up, I find that it is textured and has some weight to it. I bring it closer and smell the aroma of coffee. As I take a drink, it is hot and has a complex taste, rich and a bit acidic. I think back on buying coffee beans last week from a craft coffee roaster, and I am pleased with my choice.

In this series of events, there are many discrete experiences: seeing, touching, smelling, tasting, remembering, and pleasure. The brain is a complex assembly of separate parts, so over in one part of the brain are neural events related to one experience (such as my seeing the cup) while in another part of the brain are those neural events related to another experience (such as my remembering having purchased the coffee beans). Yet all these discrete neural events are bound together into one experience of the same object: my unified experience of enjoying a sip of coffee. As John Searle observes, "All of my experiences at present are part of one big unified conscious experience."[14]

---

13. For further development of arguments presented in this chapter, see Moreland and Rae, *Body and Soul*, 176–96; Swinburne, *The Evolution of the Soul*, 161–73. For a very sophisticated argument along these lines, see Wiggins, *Sameness and Substance*.

14. John Searle, *The Mystery of Consciousness*, 33.

Psychology, counseling, and discipleship are all based on this truth. Those who work in these fields do not help me identify and engage unrelated beliefs, desires, emotions, and choices. Rather, they study and help me correct *my* beliefs, *my* desires, *my* emotions, and *my* choices, all parts of my one big, *unified* conscious experience.

Brains Don't Have Unified Experiences

There must be something that unites these very different experiences, such that they are all said to be *my* various experiences of this one object. Physicalists say the brain accounts for this unity of consciousness. For instance, Thompson states, "An integrating, oscillating wave of electrical activity is continuously moving back and forth across the entire brain. This wave may be one way that *the brain* brings together its disparate areas into a *convergent whole*, creating our overall sense of what we feel."[15]

It is quite easy to state an opinion that any number of possible brain states cause our unified experience.[16] Yet it is much harder to give a reason to believe this is true. Neuroscientists have not found adequate data to explain our unity of consciousness. This conundrum for physicalist neuroscientists is known as the "binding problem." (This is not a problem for neuroscientists who are not physicalists, including many Christian neuroscientists, for they are not committed to reducing the human person to the brain.)

About the binding problem Searle observes, "How exactly do neurobiological processes in the brain cause consciousness? This is the most important question facing us in the biological sciences, yet it is frequently

---

15. Thompson, *Anatomy of the Soul*, 94 (emphasis added).

16. See Thompson, *Anatomy of the Soul*, 52 for another "just so" story about the brain integrating discrete phenomena. He simply states that "many experiments" have found a "correlation" between the dorsolateral (i.e., the upper and outer region) of the prefrontal cortex and attention to an object. If so, what are these experiments? And why should we reduce this correlation to identity, as discussed in chapter 3? No defense is given. Elsewhere he explains the unity of our memory due to the "emergence of the hippocampus" (a complex structure deeply embedded in the brain; *Anatomy of the Soul*, 73). Again, no data is offered in support, nor any reason to believe that even if a correspondence is shown, we should take this to mean that this unity of memory is identical to the activity of the hippocampus. Contrary to Thompson's assurances, those studying the unity of consciousness continue to see this as an unsolved problem, as discussed below.

evaded, and frequently misunderstood when not evaded."[17] Similarly, physicalist Francis Crick shares his befuddlement:

> We do not yet know how the brain puts [these various experiences] all together to provide our highly organized view of the world—that is, what we see. It seems as if the brain needs to impose some global unity on certain activities in its different parts so that the attributes of a single object—its shape, color, movement, location, and so on—are in some way brought together without at the same time confusing them with the attributes of other objects in the visual field.[18]

William Hasker summarizes, "*A person's being aware of a complex fact does not consist of parts of the person being aware of parts of the fact, nor can a complex state of consciousness exist distributed among the parts of a complex object.* Once we grasp this, we see that materialism is in deep trouble."[19]

Only a Soul Can Have Unified Experiences

To have unified experiences, we need something that is already unified and therefore can *stand under* all my many experiences. This is commonly referred to as a "*substantial* soul,"[20] which stands under and unifies all these sensations. The soul can unify these sensations because it is metaphysically simple; it is itself not composed of separable parts.[21] Therefore, it does not require something to first unify it. Rather, unified already, it can unify my discrete experiences into a whole. In my example above, it binds together these various experiences into my unified "coffee cup experience." As Aristotle states, "Discrimination between white and sweet cannot be effected by two agencies which remain separate; both the qualities discriminated must be present to something that is one and

---

17. Searle, "The Problem of Consciousness," 3; Searle, *The Mystery of Consciousness*, 60–66.

18. Crick, *The Astonishing Hypothesis*, 22.

19. Hasker, "On Behalf of Emergent Dualism," 92 (emphasis in original). See also Hasker, "Persons and the Unity of Consciousness," 175–90, or for a fuller treatment, Hasker, *The Emergent Self*, 122–46.

20. Hereafter I will typically refer to the substantial soul simply as the soul.

21. Metaphysics is the branch of philosophy that studies what is real, or the nature of things (such as the nature of the soul). In chapter 5, I will discuss in more detail the nature of the soul as simple and as a substance.

single.... What says that two things are different must be one, for sweet is different from white."[22]

The role of the soul in unifying my experiences is also known when we stop and think about what we experience. I am simply aware of myself as a soul that "stands under" and unifies all my diverse experiences. As Moreland and Rae put it, "Through introspection, you are simply aware that you ... are the self that owns and unifies your experiences at each moment of time."[23]

This awareness of ourselves as a unified self or soul that is the haver of our experiences is an intuition shared all over the world. This intuition cannot be attributed, as critics sometimes do, to Greek philosophy or religious teaching. Numerous studies demonstrate the widespread commonsense belief in a soul, even among those who have had no exposure to philosophy or religion. In his thorough study of cultures worldwide, George Peter Murdock said that ideas of the soul occur, as far as he can tell, in every culture.[24] Neuropsychologist Paul Brooks calls belief in the soul a "primordial intuition." The evidence confirms that we are "natural born soul makers, adept at extracting unobservable minds from the behavior of observable bodies, including their own."[25] Cognitive psychologist Paul Bloom arrives at the same conclusion through his research on the cognition of young children.[26]

Some physicists deny any awareness of having unified experiences grounded in their soul. David Hume, one of the greatest skeptics of the Enlightenment, argued,

> For *my* part, when *I* enter most intimately into what *I* call myself, *I* always stumble on some particular perception or other.... *I* never can catch myself at any time without a perception, and never can observe anything but perception.... *I* may venture to

---

22. Aristotle, *On the Soul*, 426b17–19, 19–20.

23. Moreland and Rae, *Body and Soul*, 183. For more on the binding problem, including physicalist responses and further critiques, see Moreland, "Substance Dualism and the Unity of Consciousness," 184–207. Similarly, Stewart Goetz (in "A Substance Dualist Response," 141) observes that the very existence of the binding problem is "confirmation of the reality of the apparent substantive simplicity of the self." I will return to this point in chapter 5.

24. Murdoc, "The Common Denominator of Cultures," 123–42.

25. Brooks, "Out of Mind," 1, quoted in Humphrey, *Soul Dust*, 95. For a good discussion of self-knowledge as good reason ("warrant") to believe that we are a soul and not a body/brain, see Plantinga, *Warrant and Proper Function*, 48–57.

26. Bloom, *Descartes Baby*; Bloom, "Religion Is Natural," 147–51.

affirm of the rest of mankind, that they are nothing but a bundle or collection of different perceptions.[27]

But *who* is entering his consciousness to poke around a bit, if not a unified self able to observe these many perceptions? And how does he know it is in fact *his* consciousness that he is poking around in? And *who* is making the affirmation that he doesn't find an enduring self? He can conclude that there is no self as the unifier of consciousness only if he first assumes that he is the owner and unifier of his experiences in the first place, to conduct this investigation.[28] To offer such an argument denying the unity of consciousness is *self-defeating*! Our unity can be shown false only if it is true and thus possible for us to be aware of our discrete perceptions.

Therefore, the fact of our unity at a particular time is one good reason, based on our everyday experience, to reject the neurotheologians' conclusion that I am identical to my brain.

Nonreductive Physicalism Is of No Help

If Thompson and Wilder are nonreductive physicalists, might they avoid the problem of our unity at any particular time? Recall that nonreductive physicalism proposes that mental events are not identical to brain events. Rather, mental events emerge from the brain (are epiphenomena), as smoke emerges from fire. In this case, Thompson and Wilder might argue that my unity is due to the mental life that emerges from the brain and not from the discrete brain events themselves.

But this doesn't solve the problem, for the same conundrum arises again at the level of these epiphenomena. In this case, discrete brain events secrete discrete mental events. Yet these mental events are no more united than the brain events that cause them. Therefore, they have no more ability to explain the unity of my consciousness than do discrete brain events. Rather than solving the problem, this move just pushes it up one level. Just naming it "emergence" doesn't in any way provide an explanation. Moreland observes, "Punting to emergence is simply to slap a label on the problem."[29]

---

27. Hume, *A Treatise of Human Nature*, I.IV.6 (emphasis added).

28. As Goetz and Taliaferro summarize, "The problem for Hume is that what he cites as evidence for [his] position . . . implies . . . that there is a substantial self that finds these things . . . not only does Hume find perceptions, but he also finds that *he* is the one who finds them." Goetz and Taliaferro, *A Brief History of the Soul*, 122.

29. Moreland, "Substance Dualism and the Unity of Consciousness," 188.

Souls Can't "Emerge" from Relationships

Another way to avoid reductive physicalism is to suggest that the soul emerges from *relationships* the brain forms. This soul, in turn, then unifies our many experiences.

At several places, Thompson and Wilder do seem to say that the soul emerges as a result of relationships with others, not from the brain *per se*. For instance, Thompson, following Daniel Siegel, states that the mind (using this term synonymously with soul) is "an embodied and relational process, emerging from and within and between brains. . . . Your sense of your mind is dependent on and shaped by your interactions with other people."[30] Elsewhere Thompson affirms emphatically, "Remember, there is no such thing as an individual brain,"[31] and "The fact that the brain responds [to other brains/persons] in such an interdependent, contingent manner reminds us that there is no such thing as a true individual."[32] Wilder agrees: "At every level of the human person, we are both created and made new by attachment love,"[33] which is a relationship between brains.[34]

This idea runs into a number of serious problems. The first is the same problem that the emergence of particular mental properties faces (as discussed in chapter 3): out of nothing, nothing comes. So out of the connections among material brains can only come other material things, rather than an immaterial soul that unifies our conscious states. Furthermore, if the soul can somehow emerge from relationships with a number of discrete brains, it is an epiphenomenon of these many relationships. Yet this cannot produce a *unified* self, but at best a group of "selves" based on the various brain relationships that come and go.

Moreover, this idea confuses what a relationship is and is not, and therefore what a relationship can and cannot do. In any relationship, there are at least three things: the two (or more) things related to the other(s), technically referred to as the relata, and the relationship *between* the relata. For instance, if I am giving you directions to a football game on Friday and say the stadium is on the west side of the high school, I am referring to three different things: (1) the high school, (2) the stadium, and

---

30. Thompson, *Anatomy of the Soul*, 29–30. See page 28 for his correlation of "mind" with "soul."
31. Thompson, *Anatomy of the Soul*, 137, 139.
32. Thompson, *Anatomy of the Soul*, 99.
33. Wilder, *Renovated*, 89.
34. Wilder, *Renovated*, 88, as discussed in chapter 1.

(3) the "west of" relation. Yet the "west of" relation exists *only because* the high school and football stadium (the two relata) *already* exist. If there is no high school and no stadium, there can be no "west of" relationship. It would be nonsense to say the relationship "west of" caused the stadium and high school to exist. To state this in more precise philosophical terms, relata are *ontologically prior* to relationships: the relata must first exist, and *only then* can relationships come to be.

Therefore, the same is true of human relationships. For any two people to be in a relationship, they must first exist. If I say "Lori and Tess are good friends," Lori and Tess must already exist in order to enter into this relationship of friendship. It would not make any sense to say, "The friendship between Lori and Tess caused Lori and Tess to exist."

It is equally nonsensical to say that "I" come to be as a result of my brain's relationship to other brains. The self cannot "emerge" *as a result* of relationships! Other people and I (the relata) would have to exist first *before* they could be in such relationships. Therefore, relationships cannot cause souls to exist. Thompson and Wilder's idea that the soul emerges from relationships is of no help in defending their position.

The inevitable conclusion is that we have a unity to our experiences that cannot be explained (1) if we are a brain, (2) if we are various epiphenomena that emerge from our brain, or (3) if we are an epiphenomenal soul that emerges from relationships with other brains. Our unity at any particular time can be explained only if we are a simple, substantial soul that unifies all our experiences. This is the first reason why the neurotheologians' conclusion that I am my brain can't be right.

Brains Cannot Explain Our Unity Through Time

I Have Unity Through Time

A second reason to reject the neurotheologians' conclusion that we are essentially a brain is our unity *through* time. Ask yourself a few simple questions: "Where was I yesterday at this time?" "Do I agree with what I am reading?" "What do I want to do this weekend?" I doubt you had any problem answering these questions concerning *your* past, *your* present, and *your* future. We know without a doubt that we exist through time—that *we* have a past, a present, and a future.

Concerning the more distant past, I keep pictures of myself in earlier years because I *know* it is really me. *I* was there! Just a moment of reflection confirms this to be the case. More recently, I know I began writing this chapter several weeks ago. I don't just think it is so, or believe it to be so. I know it to be so, because *I* was there.

The same is true of the more immediate past and present moment. Consider thinking through an argument with a series of premises that lead to a conclusion (for instance, a physicalist thinking through his argument that we do not have a soul). To come to his conclusion, he first considers the initial premise in his argument (perhaps "Everything that exists is purely physical"), and then he considers his second premise (perhaps "I exist"). Then, from these premises he draws his conclusion ("Therefore, I must be purely physical"). But drawing this conclusion is possible only *if he endures through each step in the process*, first thinking about the first premise, and then the second premise, followed by drawing his conclusion! As Thomas Reid argues, "[It] is indispensably necessary to all exercise of reason [which is] made up of successive parts.... Without the conviction that the antecedent has been seen or done by me, I could have no reason to proceed to the consequent, in any speculation."[35]

Finally, our unity through time is true of the future as well. I know that I will be present, as the same person, in a minute, an hour, a day, a week, a year, and so on. This is why I can't help but fear going to the dentist next week for a root canal. I have good reason for this fear, because it will be *me* in that chair experiencing the pain! Similarly, I am willing to sacrifice a bit now to save for retirement, for, if the Lord allows me to live that long, I know *I* will be there to enjoy it. Again, a moment's reflection confirms this to be the case—I continue to exist through all my bodily changes. I know this to be the case.

I am describing my *sameness through change*. Though I live through many changes, it is *I* who live *through* them. The language we use to refer to ourselves is based on this knowledge that we endure through time in spite of the changes in our bodies and brains. When referring to ourselves we cannot help using terms like "I" or "me" to refer to ourselves the past (for example, "I flew to Atlanta last week"), in the present (for example, "I am certainly enjoying this steak") and in the future (for example, "Please take me with you to the concert next week"). These terms are known as *indexicals*—statements from the first-person point of view that we use to describe ourselves as individuals enduring through time.[36]

---

35. Reid, *The Works of Thomas Reid*, 344.

36. This is also related to the point in chapter 3 concerning our first-person perspective, which indexicals capture in a way that is not reducible to third-person accounts.

The Brain Doesn't Have Unity Through Time

How is our sameness through change explained? Wilder believes the brain "creates and maintains a human's identity."[37] Thompson concurs:

> With the growth and neural integration of the prefrontal cortex with the hippocampus, each of us begins to develop a sense of self across time. This mental time travel, the mind's multisensory awareness of the person's past and present, along with the subsequent projection into an anticipated future, is termed autobiographical memory.[38]

However, if the brain accounts for our sameness through change, what in the brain endures through these changes? The brain, like the rest of our body, is constantly changing. Every moment, its cells are dying and being replaced by new ones. It is estimated that every seven years, all the cells in our body, including the cells in our brain, are replaced. Literally nothing is the same in the brain from our past to our present to our future, especially when we consider a time frame longer than seven years. Therefore, if I am my brain, since *my brain* constantly changes, I must conclude that *I* constantly change (or at least I am a different person every seven years, if we want to ground our identity in at least one brain cell that remains the same throughout these other changes).

Thompson's explanation of us "developing a sense of self across time . . . autobiographical memory" is of no help (even if such a thing could be produced by the brain[39]). For us to have such a sense of our self across time, there would first need *to be* a self that in fact exists across time and therefore *has* these experiences.[40]

Perhaps realizing this difficulty, elsewhere Thompson bites the bullet and agrees that since we are a brain, and since our brains are always changing, *we* are always changing—and therefore we are not to be understood as an enduring person. He states, "The mind (or brain) is fluid and always changing, if often only in imperceptible ways. In this manner, the

---

37. Wilder, *Renovated*, 68.

38. Thompson, *Anatomy of the Soul*, 74.

39. Here Thompson faces similar problems as he does in explaining our unity at a particular time, as discussed in this chapter under the heading "Brains Don't Have Unified Experiences."

40. Others have offered different criteria for identity, such as bodily continuity. As Thompson and Wilder don't endorse these options, I will not respond to these criteria here. For a discussion of these other options, see Moreland and Rae, *Body and Soul*, 180–92.

mind is never static."[41] He adds, "As far as the brain is concerned, there is in fact no such thing as the past or the future.... You can begin to respond to this 'objective reality' quite differently if you embrace the deeper reality that in some respects your past as you have viewed it doesn't even exist."[42]

Only a Soul Can Have Unity Through Time

Yet our awareness of ourselves as existing through time indicates that the physicalist can't be right—we can't be just a brain. We must be something else—something that "stands under" and lives through all these changes.[43] The only way to make sense of ourselves as living through the constant changes of our brain and body is to conclude that we are a substantial soul that *has* these experiences and life stages. "I" must ultimately be that which is present every moment of my life and makes me "me." As Dallas Willard puts it, "I am an unceasing spiritual being."[44]

Scripture affirms this conclusion. God indicates that I am the same person who was in my mother's womb—that I endure, as the same person, through the changes of my body: "Before I formed you in the womb, I knew you; before you were born, I sanctified you" (Jer 1:5). As a second example, Paul assumes he is the same person as he was before his conversion when tracing his personal history in Galatians 1:13—2:14. And as discussed in chapter 2, Scripture affirms that we will exist after death in an intermediate state and will eventually be reunited with our bodies in the final resurrection.[45]

---

41. Thompson, *Anatomy of the Soul*, 9.

42. Thompson, *Anatomy of the Soul*, 76–77.

43. For a more extended defense of the points below, as well as additional arguments, see Moreland and Craig, *Philosophical Foundations for a Christian Worldview*, 238–39 and 285–303; Moreland and Rae, *Body and Soul*, 157–96.

44. Wilder, *Renovated*, 24.

45. Some have attempted to maintain a physicalist anthropology and the reality of our final resurrection by suggesting that we cease to be when our bodies die and are then "recreated" in the final resurrection. For instance, see Van Inwagen, "I Look for the Resurrection of the Dead and the Life of the World to Come," 488–500. Such a view encounters two problems. First, it is hard to square with the biblical data (see chapter 2). Job expects to see the Lord after death, and not anyone else: "I myself will see him with my own eyes—I, and not another" (Job 19:27, NIV). Second, this view cannot explain how I can be the same person between the destruction and resurrection of my body. In what possible sense is this "me" rather than just an exact copy? (See chapter 3 for what it means for two things to be identical to one another.) As Hasker observes, "When I hope, by God's grace, to be united with my parents and other departed loved

Furthermore, neuroscience assumes this as well in the thesis of *neuroplasticity*—our discovery that we can shape our brains, or that we have the ability to "prune away those synapses that don't get much firing action."[46] This fact assumes that something must be present at each stage of this shaping. But the brain is not the same throughout the process, by definition, for it is being shaped or changed, sometimes in significant ways over time. Rather, an immaterial self or soul must be present throughout the process, overseeing this reshaping of my brain in ways I desire.

Contrary to their physicalist anthropology, Thompson and Wilder cannot help but assume they are a unified self through time. *Renovated* grew out of Wilder's being with Dallas Willard at a conference in 2012 and what he learned there. He knows *he* was there, in the past, with Willard, and bases his book, in part, on his experiences at this conference. Concerning the future, Thompson poses questions such as "What would *your* life be like if you were completely aware of the Father's deep awareness of and pleasure with you *throughout all of your waking hours*?"[47] He clearly assumes that anyone reading his book has a future—has sameness through change—in order to conceive of their future life.

Furthermore, both put their names on their books, as an affirmation that *they* had written these books. They assume that they were there every step of the way—thinking through their points, putting pen to paper, and following the process through to publication. This obviously requires them to be the same persons living through this whole process. But their brain was not there every step of the way—its molecules changed nearly every second. So Thompson and Wilder's brains did not write *Anatomy of the Soul* and *Renovated*, respectively. *They*—as immaterial selves existing through time—wrote these books (using their brains), in which they argue we are not enduring souls but fundamentally brains![48]

---

ones, it is those very individuals that I expect to meet once again—not other persons extremely similar to them, however close the resemblance might be." Hasker, "On Behalf of Emergent Dualism," 93. For a more in-depth discussion, see Davis, "Is Personal Identity Retained in the Resurrection?" 329–40. See also Loose, "Materialism Most Miserable: The Prospect for Dualist and Physicalist Accounts of Resurrection," 470–87; Kripke, *Naming and Necessity*, 113–15. For Aquinas' argument along these lines, see Pasnau, *Thomas Aquinas on Human Nature*, 391–93.

46. Thompson, *Anatomy of the Soul*, 45. He does not make the connection to the need for an enduring soul to be the "pruner."

47. Thompson, *Anatomy of the Soul*, 149 (emphasis added).

48. Similarly, consider David Hume's objection to our knowledge of ourselves at any particular time, discussed above. He first considers this perception, and then another perception, and finally concludes that he has no soul that unifies these various

Finally, Thompson and Wilder constantly use indexicals. As just one of many examples in *Renovated*, Wilder writes, "If *we* do not think about God, *we* will . . . not even notice that . . . *our* loving attachment to God is making *us* more like him."[49] Similarly, in *Anatomy* Thompson repeatedly tells stories of his clients, such as "Lydia's Story."[50] All these stories use indexicals to refer to the author and his clients as he traces their progress over time. This can be the case only if we are selves that endure through time, contrary to the physicalism they espouse.

Nonreductive Physicalism Is Again of No Help

In this case, might Thompson and Wilder be helped if they turn out to be nonreductive physicalists? Can brains produce enduring souls as epiphenomena? They cannot, for similar problems arise as we saw in relation to our unity *at* any particular time. Brain states are constantly changing. Therefore, any immaterial by-product they may produce would be constantly in flux as well. There is nothing to ground us as an enduring immaterial dimension—a soul. Nonreductive physicalism is no solution but, again, just a way to restate the problem, one level up.

In summary, denying that we are fundamentally a continuing self or soul, and claiming instead that we are fundamentally our brains, is contrary to everything we know about ourselves as continuing throughout the changes that occur in our bodies and brains. In light of this truth, both reductive and nonreductive physicalism fail. This is further reason to reject Thompson and Wilder's conclusion that we are our brains.

NEUROTHEOLOGY'S WRONG APPLICATION: HOW WE FLOURISH

We have seen why the neurotheologian's fundamental assumption (that the constant conjunction of mental and brain states means the two are identical) is wrong. From this, we have seen that their fundamental conclusion (that we are ultimately brains, or alternatively that we emerge

---

experiences. But to do so, he must first assume that his existence continues throughout the duration of his thought experiment. He must assume that he is an enduring soul in order to argue against it—a self-defeating position to be sure!

49. Wilder, *Renovated*, 49 (emphasis added).

50. Thompson, *Anatomy of the Soul*, 128–29.

from our brains) is wrong, based on the facts of our unity at any particular time and through time. So what of their application to issues of human flourishing?

Their application logically follows from their conclusion. Because Thompson and Wilder understand us as ultimately a brain, it follows that they view all problems we encounter as neurophysical problems. Accordingly, their solutions are correctives to the brain. Hence the emphasis in Thompson and Wilder's books on understanding our brain and seeking to change it. Thompson's subtitle is "*Surprising connections between neuroscience and spiritual practices that can transform your life and relationships.*" Likewise, Wilder's title, *Renovated*, proposes that this renovation is the result of us better understanding neuroscience.

Yet as we have seen, we are not fundamentally our brains. Therefore, a better understanding of our brains is not relevant to our flourishing. It is certainly interesting to know more about our brains. And there are certainly instances of neural misfirings that block some people's ability to function well (for instance, those who suffer from epilepsy). These blockages are often corrected through physical interventions such as pharmaceuticals and surgery.

However, this is not the focus of neurotheologians. They are not writing to help us understand how drugs or surgery might help us. Rather, they seek to aid our *spiritual* formation. But since, as we have established, we are not reducible to our brains, knowing more about the brain is not ultimately what helps us be spiritually transformed into the image of Christ.

Instead, we need to know more about our immaterial dimension, so that it can be formed in healthy ways (which will, given our deep functional unity with our bodies, include bodily practices). Says Dallas Willard:

> The greatest need you and I have—the greatest need of collective humanity—is *renovation of our heart*. That spiritual place within us from which our outlook, choices, and actions come has been formed by a world away from God. Now it must be transformed. Indeed, the only hope of humanity lies in the fact that, as our spiritual dimension has been *formed*, so it also can be *transformed*.[51]

What is the soul that we are to form, or the heart that we are to renovate? A correct and deeper understanding of the nature of the soul and its

---

51. Willard, *Renovation of the Heart*, 14.

relationship to the body is essential for the flourishing that neurotheologians seek—and that they will not find by following their incorrect starting assumption, conclusion, and application. In the next chapter, we will further explore the nature of the soul.

5

The True Nature of the Soul

To attain any knowledge about the soul is one of the most difficult things in the world.

—ARISTOTLE[1]

We accept that someone spends years becoming a dentist and even more years training to become a surgeon, but we do not accept that we need to spend years giving serious thought to the nature of the soul.

—DALLAS WILLARD[2]

> **CHAPTER SUMMARY**
>
> Having seen that we are ultimately a soul, we recognize that our flourishing begins with our spiritual formation. But to understand how our soul is properly formed, we must understand what it is. Here again, philosophy helps to provide clarity. The soul is an individuated human nature. A *nature* is a set of abilities, or capacities, that are true of a given type of thing. These capacities are true of

---

1. Aristotle, *On the Soul*, 402a10.
2. Johnson et al., *Dallas Willard's Study Guide to* The Divine Conspiracy, 2.

> that being, whether or not the capacities are ever lived out day by day. *Human* nature is the specific set of capacities that are true of all human beings, and only of humans. These specific capacities and their broad groupings, known as "faculties," are discussed and illustrated in this chapter, as well as how they affect one another. Finally, our human nature, which we share with all other humans, is *individuated*—that is, I am a distinct soul, with my own unique ways of expressing my human abilities. The individual soul is common referred to as a "spiritual substance." The four features of a substance are outlined, touching on some implications that will come to the fore in later chapters in relation to our flourishing.

IN THE PREVIOUS FOUR chapters, we have learned quite a bit about what we are not. We are not fully or even fundamentally our material dimension, contrary to what neurotheologians tell us. Regarding our immaterial dimension—our consciousness and soul—we have seen that the latter is the seat of our first-person perspective on life, of our reason, and of our free will. The unity of all our experiences at any given moment and our enduring through time are also due to our soul. Finally, we have seen that our soul, while ontologically distinct from our body (as indicated by its ability to live on after our body dies), is created to be a functional unity with our body. Beyond this, can we know anything else about the soul? Yes, we can.

As discussed in chapter 1, God's revelation comes in two forms. First is his special revelation: truths God has revealed in Scripture. This is the domain of theology. Second is God's general revelation: truths he has revealed in the created order. In addition to science, this includes what we can know from philosophy.

Earlier, we engaged in philosophical reflection to clarify some assumptions and draw out some logical implications of the neurotheologians' physicalism. We might say this utilizes the *negative* role philosophy can play, by showing inadequacies of various beliefs, such as those of the neurotheologians. Yet philosophy can also play a *positive* role in providing greater clarity and understanding of things we are interested in, including the nature of the soul.

Philosophers have been reasoning about the soul for millennia. Both Plato (c. 428–348 BC) and Aristotle (384–322 BC) gave this issue much thought. The conversation has been continued by countless others up to our present day. As a result, much more has been discovered about the soul.

There is certainly not complete agreement among philosophers concerning the nature of the soul. This is not surprising, as the practice of philosophy entails advancing and testing various ideas. Each idea has its advocates, who vigorously argue for their positions. Yet this fact is not true of philosophy alone—it is true of all academic disciplines, including science and theology. So the lack of agreement on a philosophical idea (or a scientific or theological idea) should not dissuade us from evaluating which position seems to make the most sense.

The good news is that a fairly detailed understanding of the nature of the soul has developed over the last two millennia. This view makes a great deal of sense and has been accepted by many throughout the centuries.

THE SOUL IS AN INDIVIDUATED HUMAN NATURE[3]

In chapter 4, we saw that the soul is *simple* in the sense of not being divisible into separable parts. However, when we begin to analyze the nature of the soul, we find that it is very complex in terms of its *metaphysical structure*.[4] It has many related properties that are structured in intricate ways. The best way to understand this complexity is to describe the soul

---

3. For a brief discussion between J. P. Moreland and me on what the soul is and why it matters, see "What Is the Soul and Why Should We Care?" episode 7 of the "Thinking Christianly" podcast (https://thinkingchristianly.org/7-what-is-the-soul-and-why-should-we-care/, accessed January 31, 2024).

4. This distinction is often described as the soul being *mereologically* simple (from the Greek word *meros*—part) and *metaphysically* complex—complex in relation to its properties. As Stewart Goetz puts it, the soul's "complexity at the level of propertyhood is compatible with simplicity at the level of thinghood." Goetz, "Substance Dualism," 37. This is an example of how making philosophical distinctions can help to untangle confusions, in some ways similar to seeing the Trinity as simple (one) in terms of nature and complex (multiple) in terms of persons.

as an individuated human nature. Let's consider what "individuated," "human," and "nature" mean, in reverse order.

What a Nature Is

To discuss what a nature is, we first must understand what properties are. Properties are specific attributes that a thing has, such as being rational. The ability to manifest a property is a "capacity" or a "disposition." For instance, if I have the ability to be rational, I have the capacity to be rational. It is a property I am able to exemplify.[5]

Furthermore, capacities come in hierarchies, from first-order to higher-order capacities. First-order capacities are those that I can exemplify right here, right now. For instance, if you ask me on the spot what 12 times 12 is, I can immediately respond, "144." I have the first-order capacity to do multiplication. But if you ask me to find the derivative of a function with respect to x right now, I will throw up my hands. I have not studied calculus, and therefore I do not have the first-order capacity to find that derivative. Yet I could enroll in a calculus course, study hard, and eventually be able to answer your question. I have the *capacity to learn* calculus. That means I have a second-order capacity that could enable me to acquire the first-order ability to find derivatives. I have the capacity to have the capacity.

But wait, there's more! To have the second-order capacity to learn calculus, I also need the third-order capacity to visually perceive mathematical numerals and equations. And this in turn requires fourth-order capacities, such as the ability to think abstractly so that I can understand what I am perceiving.

At some point, this hierarchy ends in a set of highest-order capacities. And those highest-order capacities define a thing's nature. Therefore, this is called the thing's *natural kind* or *essential nature*. Being essential entails that we cannot lose these highest-order capacities.

---

5. For more on properties, see Moreland and Craig, *Philosophical Foundations for a Christian Worldview*, 204–14. For a more detailed discussion, see Moreland, "Issues and Options in Exemplification," 133–47; Grossman, *The Existence of the World*, 14–45.

What a Human Nature Is

The essential nature of a human is a specific set of highest-order capacities shared by all and only human persons. Although other types of things share some of these capacities (for instance, angels are also rational), having this unique set of capacities is what makes a nature a *human* nature.[6]

These specific capacities can be arranged into groupings called faculties. For instance, take the capacities to reason, believe, and smell. The capacities to reason and believe are more closely related to one another than the capacity to smell. Therefore, reason and belief are grouped together in the mental faculty, while smell is located in the sensory faculty. There are six different faculties that constitute human nature.

The Mental Faculty

Our *mental faculty* includes our capacities related to apprehending and communicating goodness, truth, and beauty. This includes thoughts, which are the content of conscious states (propositions)[7] that can be expressed in sentences. It includes reason, which is the ability to relate propositions, via the laws of logic, in order to discover truth. It includes beliefs, which are those propositions we take to be true, in varying degrees of certainty. It includes the imagination, which is the ability to reflect God's image by creating. It also includes desires, which are inclinations to have or avoid certain experiences.

The Emotional Faculty

Our *emotional faculty* is our ability to experience feelings such as love, joy, grief, or anticipation. Emotional capacities differ from mental capacities (objects of thought), for we may experience emotions without thinking about them (as, for example, I experience the emotion of love when my wife suddenly enters the room). Yet we can turn our attention to an emotion and think about what we are feeling, or even consider why we

---

6. For more on "human nature," see Wallace, "In Defense of Biological Essentialism," 29–43. For a discussion of how this understanding of human nature underlies Paul's description of us, the incarnate Christ, and substitutionary atonement, see Wallace, *Aiding the Christian Scholar in Integrating Faith and Scholarship*, 42–47.

7. For more on this relationship, see Willard, "How Concepts Relate the Mind to Its Objects: The 'God's Eye View' Vindicated," 5–20.

are reacting emotionally in a certain way. At that point, our emotional and mental faculties enter into a causal relation as my emotion causes me to have certain thoughts, and my reflection may change if or how I experience the emotion.

The Volitional Faculty

Our *faculty of volition* is our ability to act freely in order to reach certain goals. This faculty also stands in interesting relations to other faculties of the soul. For example, our choices are influenced by what we believe to be true (capacities of our mental faculty) and are also usually influenced by our emotions, a fact that advertisers know and leverage all too well!

The Social Faculty

Our *social faculty* is our ability to relate to other persons. This, too, is connected in interesting ways with the other faculties. For instance, in graduate school I took a course on philosophy of mind. I remember *choosing* (volitional faculty) to pursue a relationship with Frank, a fellow student, based on my *thinking* (mental faculty) that he shared my interest in doing well in the class and my *belief* (mental faculty) that we could help one another master the content by studying together. When we helped one another get As in the class, I experienced *satisfaction* (emotional faculty).

The Spiritual Faculty

One unique aspect of our social faculty is our ability to be aware of and relate to a very special person: God. As such, it can be seen as a distinct faculty—the soul's *spiritual faculty* (again, with deep connections to the other faculties). Setting aside angels, only human nature includes this faculty. Furthermore, it is by virtue of a unique capacity in our spiritual faculty that we can have a special intimacy with God, such that we may call him "Abba," the Aramaic word for "daddy" used in Romans 8:15 and Galatians 4:6.

This ability to have such intimacy with God is an important aspect of the image of God (Gen 1:26) that all human beings bear. Granted, the Fall defaced God's image in us. That image was made inoperable, and so we were unable to express this highest-order capacity to relate to God as our "daddy" at the first-order level. However, as discussed above, since this capacity is part of our nature (a highest-level capacity), we cannot lose it completely. Though defaced, the image of God was not destroyed. As Cooper puts it, "Human nature fell when it rebelled against God. But human nature was not ontologically altered. Its created essence did not change, or so every orthodox theologian since Augustine has insisted."[8]

When we place our faith in Christ and experience his regeneration, we are able to increasingly express this highest-level capacity at a first-order level. The process of doing so is known as sanctification: becoming more and more like Christ as we are "formed in His image" (Rom 8:29; 1 Cor 15:49).

Is the Image of God Physical?

A brief comment is important here about another inroad of physicalism. Some theologians are seeking to define the image of God as a physical property, not a spiritual or immaterial property. For instance, Carmen Joy Imes argues that because the Hebrew word used in Genesis 1:26 translated "image" (*tselem*) is also used for a statue or idol found in a temple, we should interpret it physically with regard to God's image as well.[9] She states, "The imago Dei is concrete. Like a statue that represents a king

---

8. Cooper, *Body, Soul and Life Everlasting*, 199. The Fall affected not only our spiritual faculty, but every one of our six faculties. Our ability to reason, especially about the things of God, was limited; our ability to have proper emotions, especially toward God, was hindered; and so on. In this way, our depravity is "total" in its extent—it reaches to every aspect of our being. Arminians understand the doctrine of *total depravity* in this extensive sense alone, for while all faculties have been affected by the Fall, they retain some first-order ability to relate to God (retaining *some* ability to reason well about God, to have *some* proper emotions toward God, etc.). Calvinists understand total depravity *both* extensively and intensively—the Fall not only affected all faculties, but did so to such an extent that they were left inoperable in relation to God. What I'm outlining here is consistent with either understanding of the doctrine of total depravity.

9. Imes, *Being God's Image*, 31. I am not implying that Imes is a physicalist. Yet her line of reasoning is implicitly physicalist. This is similar to Thompson and Wilder's writings, which are implicitly physicalist, even if it turns out that they do not intend to endorse physicalism. In fact, in the case of Imes, she is clearly writing from what I will later define as the holistic dualist perspective, i.e., that we are a soul deeply united with our body. For instance, she states, "Our spiritual lives are embodied . . . *our* bodies are the canvas upon which *we* convey *our* sense of self to each other and the means by which we interact" (51, emphasis added). And again, "*We* are God's image. . . . no matter our intelligence or our virtue. However, *our* bodies facilitate *our* engagement with the world. (77, emphasis added). From this understanding of us as naturally embodied souls, she offers many very helpful insights throughout *Being God's Image*.

or a deity, so humans represent Yahweh to creation."[10] She argues that if the image of God is not physical, but rather identified as an immaterial property, "then we can look at people who are not performing well or who were born with limitations and conclude that they are less than human."[11]

But there are a number of significant problems with such attempts to define the image of God as physical. First, an image is not necessarily physical. Whereas a statue is a physical image, the image I have in my mind of a statue is a *mental* image, not a physical one. So there is no reason to always understand the term "image" as referring to something physical, even if sometimes it does (as in the case of statues or images in a pagan temple).

Second, if God is immaterial, it is impossible for the usage of *tselem* in Genesis 1:26 as God's image to refer to something physical. We know God is essentially immaterial, for Jesus himself says, "God is spirit" (John 4:24). Therefore, his image must also be immaterial. It is this image that he, in turn, shares with us.[12] Furthermore, we retain this image, as an essential part of our nature, after our bodies die. This is further justification for thinking that the image is immaterial.

Third, Imes's interpretation of *tselem* as only physical does not solve the problem she worries about, but actually makes it worse. Physical properties come in degrees—people have them to a greater or lesser extent. For instance, compared to others, I have more or less height, weight, and athletic ability. Therefore, if the image of God is physical, individuals necessarily bear God's image in different amounts, for each person has greater or lesser physical parts, properties, and capacities.

On the other hand, as we have seen, immaterial capacities are not degreed. We fully possess them, as part of our nature, at the ultimate-capacity level. Therefore, every human person equally and fully shares the image of God as a highest-order capacity, regardless of how fully this

---

10. Imes, *Being God's Image*, 42.

11. Imes, *Being God's Image*, 53.

12. Imes focuses on the necessity of us being physical to be God's representatives on the earth (31). However, we are not unique in this way, as the rest of creation is also physical. Consistent with her understanding of us as embodied souls, it seems she means to communicate that our embodiment is necessary for our role as God's representatives and rulers of his creation, as she states, "Ruling responsibly over the creatures God made is the way we exercise our status as God's image" (31). If this is what she means, I wholly agree. I discuss in some detail the role of the body in allowing the soul to express all its capacities, including the capacities related to bearing the image of God, in chapter 6.

capacity is expressed at the first-order level. For this reason, each of us is equally of infinite worth. Imes's worry is unfounded, and in fact, the equal and infinite value of each person can be upheld *only if* the image of God is immaterial.

Is the Spirit Something *More* Than a Faculty?

A second question arises as to whether the spirit should be taken to be a faculty of the soul. Understanding human beings to have two dimensions—soul and body, as I have outlined above—is a *dichotomist* anthropology. Yet Scripture sometimes appears to describe people as having three dimensions: body, soul, *and* spirit (most importantly in 1 Thessalonians 5:23, "May your spirit and soul and body be preserved complete," and Hebrews 4:12, "For the word of God is living and active . . . piercing as far as the division of soul and spirit"). This seems to indicate that the spirit is not one of the six faculties of the soul, but a separate entity altogether—leading to a *trichotomist* anthropology: body, soul and spirit. How are we to understand these passages that seem to indicate trichotomy?

Though this trichotomist view is often heard in popular discussions, most biblical scholars throughout church history and today have endorsed the dichotomist view, because most biblical passages indicate that we ultimately have only these two dimensions—body and soul.[13] The smaller number of passages used in support of trichotomy should therefore be interpreted in the context of this overall dichotomist picture that emerges from Scripture. References to our "spirit" are best interpreted as something real, but the aspect of the faculty of our soul that allows us to be aware of and have a relationship with God. As theologian H. D. McDonald summarizes in reference to 1 Thessalonians 5:23, "Paul is not

---

13. See chapter 2. In summarizing the overall teaching of Scripture on this matter, noted theologian A. A. Hodge flatly states that the "Scriptures teach that human nature is composed of two and only two distinct elements." Hodge, *Outlines of Theology*, 299. The trichotomist view garnered a great deal of scrutiny at the Council of Constantinople (AD 381), as Apollinarius used it as the basis for promoting a heretical view of the person of Christ that was rejected. Since then, dichotomy has been almost universally adopted. Erickson states, "Dichotomism was commonly held from the earliest period of Christian thought. Following the Council of Constantinople in 381, however, it grew in popularity to the point where it was virtually the universal belief of the church." Erickson, *Christian Theology*, 540.

giving... a scientific analysis of the structure of man's being. His concern is rather to call them to the spiritual dedication of their total lives."[14]

Furthermore, if we interpret texts that list specific aspects of the person as making ontological claims about the ultimate dimensions of a person (rather than referring to faculties of our one immaterial dimension—various parts or modes of the soul), Paul's listing of three aspects in 1 Thessalonians 5 differs from Jesus' fourfold list in Mark 10:30: "You shall love the Lord your God with all your heart, and with all your soul, and with all your mind, and with all your strength." It seems better not to interpret Jesus here as identifying a quadratic ontology of different substances or things that constitute a person (heart, soul, mind, and strength). Rather, it seems best to understand this passage, and others that list aspects of the person, as highlighting these different faculties for purposes of emphasis, within Scripture's overall dichotomist anthropology.[15]

Beyond the exegesis of specific passages and the broader theme of Scripture, the trichotomist understanding runs into the binding problem discussed in chapter 4. If we are two ontologically separate spiritual things (a soul and a spirit) rather than an ontological unity (one fundamental immaterial thing—a soul), the separate soul and spirit cannot in turn be the basis for the unity of our consciousness, for they would first need something to bind them together into a unity themselves. Yet our conscious states *are* unified. Therefore, our immaterial dimension must be ontologically unified as *one* thing, in order to cause the unity of our consciousness. As Willard puts it, "The soul is the part of the person *that integrates all the other dimensions to make one life.*"[16]

The Sensory Faculty

Last, but certainly not least, is the *sensory faculty* of our soul. This includes the capacities to smell, touch, taste, hear, and see in order to interact with the world.[17] These are our sensations—our conscious awareness

---

14. McDonald, *The Christian View of Man*, 77.

15. Erickson, *Christian Theology*, 538–43; Grudem, *Systematic Theology*, 472–82; McDonald, *The Christian View of Man*, 75–79.

16. Wilder, *Renovated*, 66 (emphasis added).

17. In chapter 3, I discussed why these are best understood as aspects of the soul and correlated to brain activity, rather than activities of the brain *per se*.

of things in the world around us, such as the feeling of the thorns on a rose bush.[18]

To experience sensations, the sensory faculty has capacities to produce a body with sensory structures that "connect" with the world and provide the neurological data for the soul to have these sensations. I realize this may sound a bit strange, as our secular age has continually promoted the idea that the body's development is a *purely* biological and physical process. Therefore, much more must be said about this faculty and its connection to the body. This discussion requires a chapter of its own. Chapter 6 will delve into this relationship between soul and body more fully.

Cause and Effect Within and Between Faculties

I have identified the six faculties of the soul and have briefly illustrated a number of ways in which capacities within a faculty relate to one another in causal ways. I also illustrated how capacities of one faculty can be related causally to those of other faculties. I will draw this all together with a more extended illustration.

First, let's consider how capacities in one faculty stand in intricate causal relations with each other. Suppose after work I decide there is just enough time to stop by Home Depot and pick up some washers to fix a leaky faucet at home. However, I must hurry, for we are having friends over for dinner and I can't be late. As I make my way to the plumbing section in the back of the store, I have the *thought* that I may have not locked my car, and that my laptop is visible on the back seat. This thought *causes* me to have another *thought*—an important report that is due tomorrow is on the laptop and not backed up anywhere else. I then *reason* that if it is stolen I would have to start writing the report all over again, and *conclude* that I would not be able to submit it on time. For these reasons I *desire* to protect the laptop, and so I *consider* returning to my car to see whether I locked it, and perhaps put the laptop in the trunk. Yet I have the *belief* that I will be in the store for only a very short time. I also *recall* that I threw my jacket in the back seat and *infer* that it probably covered the laptop. Finally, I *remember* promising my wife I'd fix the faucet tonight, as the leak is getting worse, and *calculate* that I won't have time to get the

---

18. See chapter 3 on why sensations such as a pain cannot be reduced to brain states but have a "felt" quality that is essential to them. This ability to experience the felt quality—the sensation—is a capacity of the sensory faculty of the soul.

washers if I go back to check the car. From this reasoning, I *determine* that the risk of theft is quite low and that I really need the washers today, and so I *conclude* that I should continue on to the plumbing section.

In this simple example, we see how thoughts, reasonings, desires, beliefs, and memories—all capacities within my mental faculty—are related to one another causally. (Of course, there are many more causal relationships than these few I've identified for the sake of illustration.)

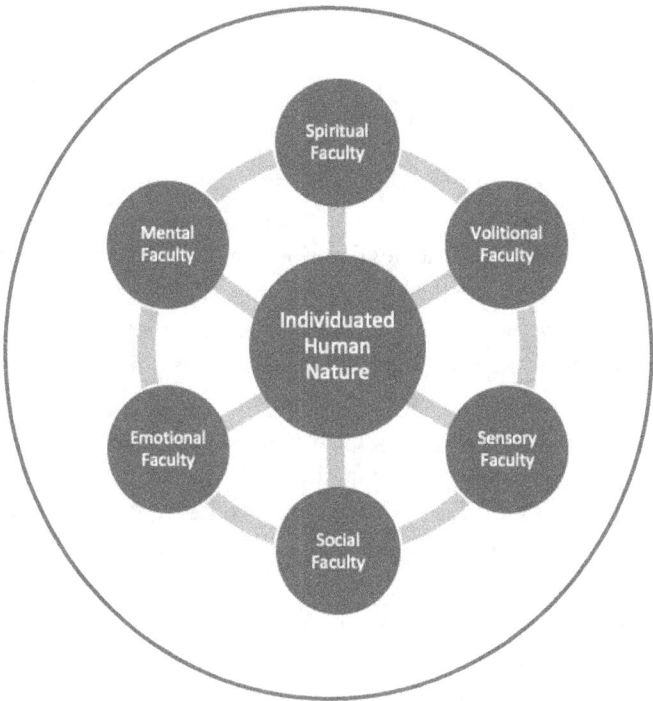

The Human Soul

Figure 1: The faculties of the soul and their causal relationships

Furthermore, let's consider how capacities in one faculty don't just operate within that faculty, but also enter into causal relations with capacities of other faculties. Continuing with this example, my thought that the car may be unlocked causes an *emotion* of panic (emotional faculty), and in response I quickly *pray* to the Lord for his help (spiritual faculty). As a result, I experience his *calming* presence (emotional faculty) and find I can again *think* clearly (mental faculty). I then *conclude* that the laptop is safe for a few minutes (mental faculty), and so I *decide* to continue on (volitional faculty). I *navigate* through the aisles to the plumbing section (sensory faculty), *talk* to the associate to be sure I get the correct washers (social faculty), *grab* the correct packet (sensory faculty), *decide* to go through self-checkout (volitional faculty) because I *think* this will save me some time (mental faculty), *scan* and *pay* for the washers and *walk* back to my car (sensory faculty). These six faculties constitute human nature. Yet one last element is needed to makes it a human *soul*: individuation.

What an Individuated Human Nature Is

Finally, our soul is an *individuated* human nature. You and I share the same human nature—we are both essentially human beings with the faculties and ultimate capacities described above. Yet at the same time, I am not you, and you are not me. We are distinct and unique human persons—*individual* human beings. For instance, we both share the faculty of emotion, but you and I express this faculty in different (individual, unique) ways. You may be better at experiencing empathy for others, and I may be better at expressing joy. Furthermore, I may never be able, at the first-order level, to experience empathy as you do, nor may you be able to experience joy as I do. The same is true of all other capacities in the other faculties, and of how they relate to one another. We are each a *distinct* person—an *individuated* human nature.

Although we know clearly *that* we are individual persons, it is not clear *how* this is so, and debate continues on this matter. However, we don't need to know *how* something is the case to know *that* it is indeed the case. This is not only true in philosophy. For instance, in theology we know *that* God is a triune being, though we don't know *how* this is so. In the same sense, we know *that* we are individuals though we don't fully understand *how* this is the case.[19]

---

19. For a summary of the issue of individuation, see Audi (ed.), *The Cambridge Dictionary of Philosophy*, "Individuation," 367–68. For a plausible defense of "bare particulars" as the cause of individuation, see Moreland, "Theories of Individuation: A Reconsideration of Bare Particulars," 251–63.

AN INDIVIDUATED HUMAN NATURE IS A SPIRITUAL "SUBSTANCE"

We have seen that the soul is a certain type of thing (a nature) with the specific highest-order capacities that make it a human type of thing (a human nature), and that it is individuated so as to make us unique persons. As such, the soul is what we call a *substance*. It is a *substantial soul*.

It is crucial to have a proper understanding of our souls as substances. As Willard writes in *The Divine Conspiracy*, "To understand spirit as 'substance' is of the utmost importance in our current world, which is so largely devoted to the ultimacy of matter. It means that spirit is something that exists in its own right."[20] He defines a substance as "an individual thing that has properties and dispositions natural to it (i.e., as part of its essence), endures through time and change, and receives and exercises causal influences on other things."[21] As Willard indicates, the word "substance" as used in philosophy is not what we now commonly understand the word to mean. We usually use "substance" to mean something physical. But in philosophy, its meaning is quite different. To understand what a substance is, and to therefore understand the soul as a spiritual substance, we must outline the four defining features to which Willard refers.[22]

Substances Are Owners and Unifiers of Properties

First, as the name indicates, a substance "stands under" other things. In Willard's terms, it "has properties and dispositions." As discussed earlier in this chapter, people have many properties, such as being kind or intelligent. Yet these properties are not just floating around. They are always attributes of some *thing*. That thing is a substance—a specific person who has and unifies these properties. Of this, Willard says in *Renovated*, "The basic function of the soul is to put all these parts together and make one life out of it."[23] Yet substances are not in turn possessed by anything more

---

20. Willard, *Divine Conspiracy*, 82.
21. Willard, *The Great Omission*, 138.
22. For more on what a substance is, see Moreland and Craig, *Philosophical Foundations for a Christian Worldview*, 214–26. For a more detailed treatment, see Hoffman and Rosenkrantz, *Substance: Its Nature and Existence*. For a discussion of rival views, see Loux, *Substance and Attribute*.
23. Wilder, *Renovated*, 57.

basic—in other words, nothing "stands under" a substance. God created substances to be the bedrock of all other reality.

Substances Are Enduring Continuants

Substances can gain and lose parts and properties over time and yet remain the same. In the terminology of chapter 4, substances have sameness through change, or as Willard says, they "endure through time." For instance, a person can increasingly exemplify the property of wisdom while remaining the same person, as this highest-order capacity is increasingly expressed at the first-order level. Or a person can gain and lose physical parts (such as assimilating new atoms in the brain or having a leg amputated) and remain the same person. Furthermore, as discussed in chapter 2, this substantial soul endures after the destruction of the body and is reunited with the body at the final resurrection. As Willard says, "We may be sure that our life—yes, that familiar one we are each so well acquainted with—will never stop. We should be anticipating what we will be doing three hundred or a thousand or ten thousand years from now . . . we are never-ceasing spiritual beings."[24]

Substances Change in Law-Like Ways

Substances undergo change in the first-order exemplification of properties in predefined ways, according to the nature of the specific substance. For any capacity discussed earlier (such as reason), there are specific developmental pathways that all who share human nature *naturally* go through. Thus, substances have a *teleology* (from the Greek word *telos*, meaning "end") or "goal" toward which they naturally develop. According to Willard, they "exercise causal influences" toward this telos. This end is the expression of *all* highest-order capacities at the first-order level.

As this occurs, we are said to be *maturing*—thinking, handling emotions, and relating to others in more mature ways. The reality of this law-like change toward a predefined end in various areas of human development allows professionals to chart a person's growth (mentally, physically, emotionally, and socially). In light of this information, they can determine whether a disability is present and, if so, how to intervene.

---

24. Willard, *Divine Conspiracy*, 86.

For instance, if a three-year-old cannot write a coherent sentence, we all understand that this is normal for his age. But if this is still the case at age 6, we know there is a deficiency in mental capacities being expressed and intervene with specialized educational programs accordingly. Emotionally, when a two-year-old throws a temper tantrum, we say it's just the "terrible twos," but if it happens at age 15, we call it a deficiency in emotional development that must be corrected. Volitionally, we don't fault a seven-year-old for being unable to make wise decisions concerning their finances, but we do blame someone who can't do so at age 27. As Willard summarizes, "Maturity refers to a process of growth. Immaturity stands at one end of that process, maturity at the other. And the growth process that results in maturity involves the essential components of the human being and some of the dynamics of their interaction."[25]

Substances Are Particular Things

Finally, a substance is "an individual thing": *this* dog Cozette, *this* angelic person Gabriel, or *this* human person Beth.[26] It is impossible for something to be less than one thing. Yet each of us is not more than one thing either, as our discussion in chapter 4 of the soul's mereological simplicity (unity) showed. Each of us is one and only one particular thing. Willard assumes this throughout his writings as he regularly describes how each of us, as individuals created in God's image, can be spiritually formed so as to flourish in his Kingdom now and forevermore, based on the fact that "*You* are an unceasing spiritual being *with an eternal destiny.*"[27]

Taking this all together answers the question raised at the beginning of this chapter: What is the soul? It is an *individuated human nature* or an *immaterial substance*. But more must be said about the soul's relation to the body. We turn to that question next.

---

25. Wilder, *Renovated*, 54.

26. This is what Aristotle defines as a primary substance, as discussed in his *Categories* and *Metaphysics*. This is not to be confused with what he terms "secondary substances," which refer to the *type* of thing a primary substance is (its natural kind or essence, such as humanness, discussed above).

27. Wilder, *Renovated*, 24 (emphasis added).

6

The Unity of the Soul and the Body

A cheerful heart is good medicine, but a broken spirit saps a person's strength.
—PROVERBS 17:22 (NLT)

Matter, ordinary physical "stuff," is the place for the development and manifestation of finite personalities who, in their bodies, have significant resources either to oppose God or to serve him.
—DALLAS WILLARD[1]

> **CHAPTER SUMMARY**
>
> Since our souls are deeply united with our bodies, we must understand the relationship between the two if we want to properly understand spiritual formation and human flourishing. This chapter explains the connection between the soul and body, beginning with the soul's formation of our material dimension and carrying on through its natural development of our bodies. This discussion helps us understand normative concepts like *disease*, *dis*abilities, and age-*appropriate* development. The ways

---

1. Willard, *Divine Conspiracy*, 254.

> in which our bodies and souls have effects on one another are also discussed, highlighting the causal connections between the two. An example of this relationship is taken from Dallas Willard to illustrate the relation of soul and body. This understanding is consistent with what biblical studies and philosophy tell us we are, and it also makes sense of what we are learning about the brain from neuroscience. Finally, terms used for this view are discussed, with holistic dualism being my preferred name due to its clarity and descriptiveness.

IN CHAPTER 5, WE saw that we are substantial, immaterial souls. However, we are also *embodied* souls. This incarnation is so deep because the body is the result of the soul's faculty of sensation. This accounts for our functional unity. How should we understand this functional relationship between body and soul? In contrast to the physicalism of the neurotheologians, this chapter will consider a robust depiction of the importance of the body *in relation to* the soul. The interpretation presented here is upheld not only by Scripture (as shown in chapter 2), but in various ways by many important figures in church history such as Thomas Aquinas, John Calvin, C. S. Lewis, and Dallas Willard.[2]

BODY AND SOUL: A MATCH MADE IN HEAVEN

As discussed in the previous chapter, each faculty of the soul contains highest-order capacities that are naturally, over time and if not impeded,

---

2. The anthropology of Aquinas will be discussed throughout this chapter. Although Calvin certainly embraced the broad contours of this view, the extent to which he shared the details of the view presented here is debated. For a good treatment of how Calvin and other Reformers applied Aristotelian thinking, see Helm, *Human Nature from Calvin to Edwards*. For C. S. Lewis's anthropology, see his *Till We Have Faces: A Myth Retold* and Taylor's helpful summary of Lewis's anthropology in his, "In His Image and Into His Likeness: Human Nature's Theosis in C. S. Lewis's *Till We Have Faces: A Myth Retold*." Willard's anthropology is discussed in some detail in this chapter, as well as in chapter 7.

expressed at the first-order level. This allows us to fully express our nature and thus live as we were created to live.

For instance, we have highest-order capacities to experience the love of God and others, and in turn to express love. As we live out these capacities at the first-order level, we are living most fully according to our nature. One way we often describe this concept is that we are "flourishing." As plants flourish when they are able to fully blossom, in accordance with their nature, so too we are flourishing when we can fully "blossom" emotionally and experience healthy emotions according to God's design. The same is true of all our other faculties: we flourish as we are able to fully express those capacities at the first-order level.

God has created us with sensory capacities to be aware of and engage the physical world around us. These are the capacities to see, hear, smell, taste, touch, and experience other physical sensations such as a pain or an itch. These capacities allow us to move about, speak to others, avoid danger, create artifacts, and so on. Ultimately, these capacities enable us to live well and flourish in this world as we accomplish the ends for which God created us.

Yet we need a body in order to exercise these sensory capacities to these ends. As our soul comes to be, it immediately begins to form the physical properties needed for this purpose (physical structures such as noses, ears, and so on—collectively the "body"). One plausible explanation of how the soul does this is through its structuring of DNA.[3]

Then, as the body develops, it in turn begins to provide the soul with experiences of the world as it accesses the physical realm through these organs. This can be illustrated by the relationship between sounds (music) and a CD. Recording equipment accesses sounds, correlates them with digital sequences, and then "stores" these sequences in the CD's grooves. Note that the music itself is not in the CD, but only the correlated digital sequences. A proper retrieval system (in this case, a

---

3. Richard Connell has made a compelling case as to how the soul arranges molecules to shape DNA in order to form the body necessary to obtain its ends. See Connell, *Substance and Modern Science*, 89–118, 185–201. For a good summary and additional sources, see Moreland, "In Defense of Thomistic-Like Dualism," 107–8; Moreland and Rae, "Substance Dualism and the Body: Heredity, DNA and the Soul," chapter 6 of *Body and Soul*, 199–228. This view is most consistent with the theological position on the origin of the soul known as *traducianism*, which holds that the soul comes to be at the time of conception, transmitted in the act of reproduction. Its theological merits (over the alternative view that God directly creates each soul and "places" it in a body) include accounting for the transmittal of Adam's sin nature without God being the direct creator of sin, and the fact that God rested from all acts of creation after the sixth day. For more see Elwell, "Traducianism," 1106; Hodge, *Systematic Theology*, 65–78.

CD player) is necessary to access these digital sequences for the correlated sounds to be heard. Without a CD player retrieving these digital sequences, there is no music.

In a similar way, our ears access sounds and correlate them with neural impulses that are then "stored" in the brain. Note that the sound itself is not in the brain, but only the correlated neural sequences. The necessary retrieval system—the soul, through the faculty of sensation—is necessary to access this neural "raw material" and experience the intrinsic, irreducible "felt quality" of these sounds.[4] Without the soul doing its job, there is no experience of the sound.

The same is true of the relationship between our other sensory organs and our soul's experiences: of the nose to my soul experiencing the smell of my wife's perfume, the mouth to my soul experiencing the taste of an orange, the eyes to my soul experiencing the sight of a sunset, and the skin to my soul experiencing the roughness of sandpaper (as well as the feelings, thoughts, and information these experiences result in, due to the causal relations between the various faculties of the soul).[5]

As is true of capacities in other faculties, it takes time for all these highest-order sensory capacities to be expressed at the first-order level. First, all the physical structures composing the body must fully develop. This process of physical maturation begins in the womb, as we start to develop the physical structures necessary for our souls to experience sensations. This process continues after birth until we reach full physical maturity. At that point, our bodily structures allow our souls to fully engage the world as we were designed to do, so that we can flourish. As Des Chene puts it, "The human soul is not merely joined with the body in fact. It is the *kind* of soul which ... by its nature presupposes union with a body, and moreover, with a particular kind of body, a body with organs, in order to exercise all its powers."[6]

---

4. This is the *phenomenal* "what-it-is-like-to-hear" quality of sounds. Refer to chapter 3 for reasons why neural events are not identical to the soul's experiences of hearing sounds, feeling pains, and so on.

5. For instance: the feel of rough-cut lumber and the smell of sawdust causing me to think of my dad, the hearing of "Hark the Herald Angels Sing" causing me to experience worship, the sight of my children causing me to experience love, the taste of Skyline chili causing me to remember my days in high school, and reading or hearing the sentences "Brooke is older than Ryan" and "Ryan is older than Luke," causing me to reason that "Brooke must be older than Luke."

6. Des Chene, *Life's Form*, 71. For more on how Aquinas expressed this, see Moreland and Wallace, "Aquinas vs. Locke and Descartes on the Human Person and End-of-Life Ethics," 319–30.

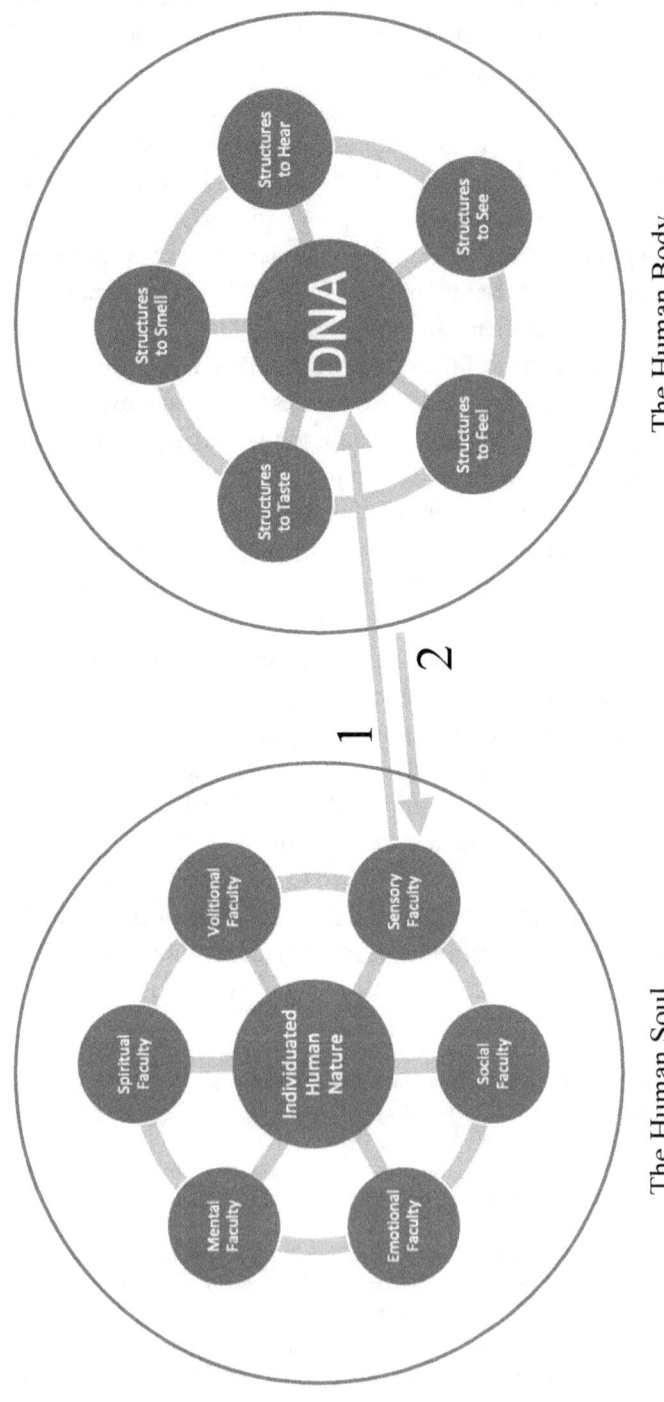

The Human Soul　　　　　　　　　　The Human Body

Figure 2: The soul, body, and their causal relations (1 and 2 sequentially)

However, in some cases, higher-order blockages hinder these physical structures from developing appropriately, which means that we cannot exemplify these sensory capacities at the first-order level. This, in turn, hinders our flourishing. Again, this is similar to a CD player being defective, such as having a faulty power supply. Its laser still has the ability to translate the digital sequences into sounds, but it can't do so because of this defect. As a result, the CD player is hindered in doing what it exists to do—play music.

To remove these blockages, professionals in various functional areas have developed means of intervention (much as those with training in electronics can fix a CD player's power supply). For instance, a child's facial muscles may not develop so as to shape his mouth appropriately to pronounce words. Intervention may include surgery to correct the muscular attachments, physical therapy to strengthen the proper muscles, speech therapy to train the muscles to move in the proper ways, or some combination of these. Once this higher-order deficiency (the muscular limitation) is addressed, the impediment is removed and the child can express the first-order ability of making the sounds necessary to articulate words. To take another example, areas of the brain used by the soul's mental faculty may not develop appropriately. As a result, the person cannot express mental capacities at the first-order level. Neurosurgeons may attempt to correct this limitation in the brain through a specific surgical procedure, or neurologists may prescribe medications. If these interventions are successful, the brain begins functioning properly, and the soul can now express its mental capacities at the first-order level by reasoning well.

Furthermore, like capacities in other faculties, these sensory capacities are related to one another and to capacities of other faculties. For example, suppose you hear a sound, such as the sentence "Let's go get ice cream." This first-order *sensory* experience (an experience of your soul, via the neural pathways from your ears to your brain) causes a number of *mental* events to occur: the *thought* that you like ice cream, the *desire* to enjoy ice cream with your friend, and the *belief* that you can have ice cream today without blowing your diet. All this, in turn, causes you to *decide*, using your *volitional* faculty, to accept her offer to get ice cream, which in turn causes you to use your breath, vocal cords, and the shape of your mouth to express the semantic content of the decision you have made by *speaking* the words, "Sure, let's go," and by beginning to *move your feet* in the right direction.

Notice in this example how your soul uses your body, including your brain, to function effectively in the world—in this case, to enjoy God's great gift of ice cream with a friend. The causal chain first runs from your body to your soul as your body provides the neural events necessary for sensation, causing you to hear the words, understand them as an invitation, and finally decide to accept. At that point, the causal chain runs back from soul to body as you expel breath over your vocal cords, form your mouth in certain ways to speak words of acceptance, and direct your steps toward the ice cream shop.

Such two-way causal connections between the body and soul occur in all our interactions with the world. If I choose to worry, I may develop an ulcer. If I have a poor diet, I will find it harder to have coherent thoughts and proper emotions. Even my fingers pressing computer keys to type this paragraph illustrate my ability to take ideas formed in my mind and then use this part of my body to type the words that express these ideas. As you read these sentences using your eyes, in turn your mind will understand the propositions being communicated in these sentences, as well as the logical connections among these propositions, and eventually (I hope) decide that you agree with my conclusions. As Swinburne summarizes, "Our bodies are the vehicles of our knowledge and operation. The 'linking' of the body and soul consists in there being a body which is related to the soul in this way."[7]

Following Aristotle, Thomas Aquinas (1225–1274) observed this causal relation from soul to body—that the body results from the soul[8]—long before the discovery of the critical role DNA plays in our bodies' formation. Aquinas's view can be summarized in this way: "A soul is the form that organizes material components into a living organism with its internal organs, limbs, and so on and makes that organism capable of human action, both bodily and mental."[9]

Much more recently, this relation of soul and body is what Dallas Willard was referring to when he spoke of the body as our "powerpack"[10] that allows us, as a substantial soul or a person, to engage with the world. He summarized this point clearly when speaking of spiritual disciplines: "*You* engage *your body* in it."[11] He also illustrated this relation with the

---

7. Swinburne, *Evolution of the Soul*, 146.
8. Aquinas, *On Spiritual Creatures*, IV ad 9.
9. Goetz and Taliaferro, *Brief History of the Soul*, 50.
10. Wilder, *Renovated*, 13.
11. Wilder, *Renovated*, 64 (emphasis added).

analogy of a driver and an earth mover: the driver uses the earth mover to engage the world (move the earth) in the same way as a person uses his or her body to engage the world (interact with the physical world in desired ways).[12] In *The Divine Conspiracy*, he clarifies this relationship: "Matter, ordinary physical 'stuff,' is the place for the development and manifestation of finite personalities who, in their bodies, have significant resources either to oppose God or to serve him."[13]

This is expressed in various ways. We may say the soul "enlivens" the body—makes it alive. Or we may say that it "animates" the body (from the Latin *animus*, "soul" or "life force"). It is only when the body is ensouled that it has this internal unity necessary for its parts to work together. Only then is it alive or animated. If it is no longer ensouled, it is only a corpse that *used to be* a human body (*used to be* alive or animated). This fact will become obvious soon enough, as the body begins to decompose without life from the soul. As Aristotle says, "A dead man is a man only in name."[14]

The soul enlivens the body by being *in* the body. However, the soul is not in the body in a spatially extended way, as if a little piece of my soul were in my finger and another little piece of my soul in my knee. If this were the case, since I am 6'5" tall, I would have much more soul than my 5'6" wife. But this is certainly not true! Or if someone has a leg amputated, he would literally lose a part of his soul. But neither is this the case. As Augustine says, we must not think of the soul as "diffused throughout the whole body, as is the blood."[15] Rather, the soul is present throughout the entire body non-spatially, maintaining the body in existence, in the same way God is present throughout the entire world non-spatially, maintaining the world in existence.[16] This understanding of the relationship between body and soul makes sense of our experiences living in the world, as illustrated in various ways above.

---

12. Wilder, *Renovated*, 59–60. More will be said about this analogy in chapter 7.

13. Willard, *Divine Conspiracy*, 254.

14. Aristotle, *Meteorology* IV.12: 389b31–32. Elsewhere he says that a human hand, when severed from the body, is no longer a human hand (*Metaphysics* Z11, 1036b25–33 and *Meteorology* IV.12: 389b31–32). This idea is akin to James's words as he compares the relation of soul and body with faith and works: "faith without action is as dead as a body without a soul" (James 2:26, Phillips).

15. Augustine, *Greatness of the Soul*, 30.61. For more on Augustine's argument in support of this point, see Goetz and Taliaferro, *Brief History of the Soul*, 43–45.

16. See Goetz and Taliaferro, *Brief History of the Soul*, 43–45, 140–46; Moreland and Rae, *Body and Soul*, 199–228.

UNIFYING THE DISCOVERIES OF THEOLOGY, PHILOSOPHY, AND NEUROSCIENCE

This understanding of the relationship of soul to body is also most consistent with what we know from theology, philosophy, and neuroscience. From theology, the Creation account in Genesis 1 makes clear that all of God's creation is good. This includes our creation as embodied creatures with a functional unity. This was God's idea, and by his design we best flourish in this unity. Proverbs 17:22 illustrates this functional unity: "A cheerful heart is good medicine, but a broken spirit saps a person's strength" (NLT).

Yet, as we have also seen from Scripture, the soul and body are an ontological duality, due to the fact, among other things, that we can live apart from our bodies. This makes sense of what we are said to experience during our intermediate, disembodied state. During that period, without our brains, we can still think, reason, desire, and feel emotion.[17] Concerning the biblical data Cooper summarizes, "This is a view of the human constitution which flows out of Scripture, preserves the heart of the church's historic doctrines, and can provide a framework for respectable participation in the modern intellectual world."[18]

This understanding of what we are as a soul deeply united with our bodies is also most consistent with what we have discovered from philosophy, as outlined in chapters 3 and 4. As a substantial soul, we unify all of our various faculties' experiences. As a substantial soul, we also continue to be the *same* person who has these successive experiences of our various faculties moment by moment, day by day, week by week, and year by year.

Finally, as discussed in chapter 1, neuroscientists have discovered correspondences between brain events and mental events, as well as our brain's neuroplasticity. Thompson and Wilder take these two facts as

---

17. Recall from chapter 2 the discussion of Jesus' promise to the thief dying on the cross next to him: "Today you will be with me in paradise" (Luke 23:43). This promise assumes that after death the thief would have an awareness of being with Jesus, and that he would be able to reason along the lines of "While on the cross I believed in Jesus as the Son of God, and he promised me we would be in paradise later today, so that's where I must be now." It is also fair to assume that the thief would experience joy as a result of this reasoning and, in turn, desire to praise God. In sum, he will still have a mental life, though there will not be neural activity. In some way, God sustains us in this unnatural state until our bodies are resurrected. Here again we know *that* this is the case from Scripture, yet we do not know *how* he does this.

18. Cooper, *Body, Soul, and Life Everlasting*, 5.

reasons to see us as fundamentally physical beings. However, these facts are better explained by understanding our souls and bodies as functionally united but ontologically distinct.

Concerning the correspondence between brain events and mental events, we have already seen why this constant conjunction does not mean that the two are identical. The understanding outlined in this chapter helps us understand and in fact anticipate why neural and brain events occur together: any time we engage the world the body and the soul are *both* involved, causally affecting one another. As such, they form a functional unity that connects them in profound ways. Yet they are not identical, for the body is itself the expression of a range of powers in the soul needed for us to flourish in the world through a body. As Moreland summarizes, "Since the soul . . . forms the body to express its powers that depend on certain bodily structures, the soul unfolds a brain suited for its faculty of mind to be fully operative. Brain complexity and mental sophistication correlate because the brain was formed to express mental functioning."[19]

As for the brain's neuroplasticity, what is the cause of the brain being reshaped? The cause will involve reason and choice. There must be a rationale to reshape the brain. Based on these reasons, a choice must be made to do so. As discussed in chapter 3, the brain lacks the ability to reason or choose. Therefore, the brain cannot reshape itself. However, a substantial soul deeply united to the body (including the brain) has this ability to reason and to choose. It follows that the soul, and only the soul, can account for the reality of the brain being reshaped. My soul is what creates new neural pathways that allow me to flourish more fully.

Beyond this, other findings of neuroscience are best explained by this account of the relation between soul and body. For instance, it explains cases such as Dandy-Walker Syndrome, in which important parts of the brain do not develop. In other cases, the *corpus callosum* (a large bundle of nerve fibers that connect the two hemispheres) does not develop at all. Yet these individuals can still think, feel, choose, and desire. If the brain itself is the cause of these abilities, as physicalists maintain, this is unexplainable. Yet if the soul uses the brain to do these things, it makes sense that the soul can adapt and compensate in order to still accomplish its ends.

Other studies have discovered that a connection between our brain's two hemispheres, though normally the case, is not *necessary* for our mental life. Nobel laureate Roger Sperry has done research on patients

---

19. Moreland, "In Defense of a Thomistic-Like Dualism," 114.

with epilepsy that involves cutting the corpus callosum, in some cases almost entirely, so as to sever the brain's two hemispheres. After surgery, these patients experience fewer seizures. Yet they experience no other appreciable differences from those with fully interacting hemispheres.[20] They continue to experience an integrated mental life (one stream of consciousness, rather than two separate streams from the two disconnected halves of the brain). They also continue to appear to others—including friends and family who know them very well—as having the same integrated sense of self, with the same personality and abilities. Again, all this makes sense only if persons are substantial souls that can use and even (if necessary) repurpose their brains.

There are also cases of *hemispherectomies* (the surgical removal of one side of the brain) after which patients can still process language, feel sensations, and move both sides of their bodies, no matter which side of the brain has been removed. Again, if the brain is the ultimate cause of these activities, as physicalists believe, this should not be possible. But if the soul uses the brain to achieve its ends, it makes sense that the soul can adapt how it uses the brain to reach its ends in these unique cases.

Finally, this understanding makes sense of research identifying the limitations of medication in treating emotional disorders. Medications are often of great help, as emotional challenges can be related to neurological disorders (again, cases of blockages in the brain resulting in the person not being able to exemplify highest-order emotional capacities at the first-order level). However, if we are only physical, pharmaceuticals would, by definition, *completely* address all emotional challenges. Yet this is not the case. For instance, take the case of Mike, who has Social Anxiety Disorder (SAD), resulting in excessive fear of being with others. The

---

20. After extensive research, Sperry discovered that patients experienced only very minor perceptual limitations. Yet it took Sperry, a trained neuroscientist, days of interacting with the patients to identify these differences. What is remarkable is what Sperry *did not* find. He found no limitation in the patient's ability to do mathematics, discuss virtues such as justice and mercy, or engage in any other abstract reasoning. For more on Sperry's research, see Michael Matheson's interview with Michael Egnor, M.D., Professor of Neurological Surgery and Pediatrics and Director of Pediatric Neurosurgery at Stony Brook University's School of Medicine, on Episode 13 of *The Moral Imagination Podcast*, "Are We Our Brains? Philosophy and the Foundations of Neuroscience," October 21, 2020, https://www.themoralimagination.com/episodes/michael-egnor, accessed January 31, 2024. His discussion of Sperry's work begins at the 52-minute mark. Egnor adds that if a heart was cut in two and blood continued to be pumped, we would have to rethink what we understand the heart to be doing. The same is true concerning the brain as we find that it can be cut in two and yet we continue to function in virtually the same way.

most commonly used treatment for SAD has been medications. However, this has not been entirely effective. Research has now confirmed that non-medical treatments such as talk therapy are more effective.[21]

In sum, if one examines the data from neuroscience without a prior commitment to physicalism, one will conclude that the mind and brain are distinct yet also maintain deep causal connections to one another.[22] As "all truth is God's truth" and Scripture paints a picture of us as a deeply united body and soul, we should expect this to be confirmed by general revelation—in this case, by philosophy and neuroscience. Cooper summarizes well the interplay of theology, philosophy, and science on the nature of what we are, saying that Christian believers committed to the biblical view of our body and soul as a functional unity and ontological duality can "work out of their commitment and belief as scientists and philosophers with integrity. There is no need for pitting the truths of revelation against the current results of reason. One need not pick between biblical orthodoxy and an adequate theoretical model of human nature."[23]

NAMING THIS VIEW: "HOLISTIC DUALISM"

This understanding of what we are has two essential features. First, it recognizes us as an ontological duality that includes an immaterial substantial soul and a physical body. Views of this sort are designated as *substance dualism*. This term explicitly names the reality of our ontological duality.

---

21. Society of Clinical Psychology, "Case Study Mike (Social Anxiety)." This case and other research are summarized in Johns Hopkins Bloomberg School of Public Health, "Talk Therapy—Not Medication—Best for Social Anxiety Disorder, Large Study Finds." Interestingly, Wilder discusses similar cases (for example, in *Renovated*, 122–24), as does Thompson (for example, in *Anatomy of the Soul*, 53–57), but apparently without realizing that these case studies are more consistent with the view I'm offering than with their physicalism.

22. For more on this, see the groundbreaking work by Penfield, *Mystery of the Mind*. He concludes "that it is easier to rationalize man's being on the basis of two elements than on the basis of one" (114). For observations along these lines from another world-class neuroscientist who received a Nobel Prize for his work in brain physiology, see Eccles, *Facing Reality*. Similarly, see also Popper and Eccles, *The Self and Its Brain*; Eccles and Robinson, *The Wonder of Being Human*; UCLA neuroscientist Schwartz and Begley, *The Mind and the Brain*; Beauregard and O'Leary, *The Spiritual Brain*.

This is not to deny the brain can also be affected by external, material causes. Pharmaceuticals can interact with the brain. Neural implants are being developed to interact with the brain. Yet these external influences do not count against the deep unity of mind and brain, but only add an additional element in the causal relations.

23. Cooper, *Body, Soul, and Life Everlasting*, 230.

However, there are two main varieties of substance dualism. The second essential feature of the view I'm espousing understands our soul as deeply united with our body, forming a functional unity, which is necessary for our full flourishing. This distinguishes the position I've outlined from the other main variety of substance dualism—namely, *Cartesian dualism* (to be discussed in chapter 8). A number of titles have been used for the form of substance dualism I am endorsing.

Due to its clarity and descriptiveness, I prefer the designation that Cooper offers: "All things considered, therefore, the biblical view of the human constitution is some kind of *'holistic dualism.'*"[24] It is dualistic in that it affirms we have two irreducible dimensions,[25] and holistic in that these dimensions work together in harmony. Cooper continues, "My definition of dualism certainly does not imply the antithesis and essential disharmony between body and spirit"[26] and "The dualism of biblical anthropology presented here is part of the same picture which yields functional holism."[27] Yet our soul is ultimately what we are, as it is what continues in the disembodied state: "It is clear that the doctrine of the intermediate state logically requires the possibility that persons can exist without earthly bodies.... The view of the human constitution implied by the doctrine may be labeled 'dualism' or 'ontic duality.'"[28] In summarizing his in-depth study of all biblical texts related to anthropology, Cooper concludes, "Since there are no other unanswerable challenges to it, holistic dualism ought to be embraced by Christians without reservation."[29]

However, some use the term "holistic dualism" to refer to us not as a *functional* unity, but rather as an *ontological* unity.[30] For the reasons

---

24. Cooper, *Body, Soul, and Life Everlasting*, xvi (emphasis added).

25. Technically, per the definition of a substance as an enduring continuant as discussed in chapter 5, the body is not a substance. It is better to call the body an ensouled physical structure, or a mode of the soul, as it has the physical properties necessary to express the soul's sensory capacities. However, the view is dualistic in the sense that the material and immaterial dimensions are two distinct aspects of us as human persons.

26. Cooper, *Body, Soul, and Life Everlasting*, 163–64. In this, he is distinguishing his view from the Cartesian understanding of the soul-body relation to be discussed in chapter 8.

27. Cooper, *Body, Soul, and Life Everlasting*, 164.

28. Cooper, *Body, Soul, and Life Everlasting*, 164. For his more complete summary of holistic dualism, see pages 160–64 and 194–96. See pages 187–98 for why he prefers ontological dualism over mere functional dualism.

29. Cooper, *Body, Soul, and Life Everlasting*, 231.

30. For instance, Green states, "Holistic dualism . . . posits that the human person, though composed of discrete elements, is nonetheless to be identified with the whole." Green, *In Search of the Soul*, 13.

discussed in previous chapters,[31] this is *not* the way Cooper or I are using the term *holistic dualism*.

Equally apt monikers identify this view with key features in the anthropology of Thomas Aquinas. He thought much about what we are, drawing heavily on the important insight from Aristotle that individual things in the world are composed of *both* form and matter. By this, he meant that things in this world certainly *are* physical, yet are not *purely* physical. All physical things are "formed" as they are as a result of the immaterial form (essence) which causes the matter to be what it is. The form and matter are deeply united in each particular thing.[32] Therefore he says, "The soul *plus* the body constitute the animal."[33] This is known as Aristotle's doctrine of *hylomorphism* (from the Greek *hulê*, "matter," and *morphê*, "form"). Aquinas followed Aristotle in this regard, saying, "We must not think, therefore, of the soul and body as though the body has its own form making it a body, to which a soul is super-added, making it a living body: but rather that the body gets its being and its life from the soul."[34] Again, the soul is that which animates the body.

Due to Aquinas's use of Aristotle's hylomorphism to understand us as a deep unity of soul and body, Rob Koons identifies this view as *Thomistic hylomorphism*.[35] Similarly, J. P. Moreland terms this view *Thomistic-like dualism*.[36] The "like" in the title identifies some modifications of Aquinas's specific understanding. Yet he states, "Our view shares enough of the important aspects of a Thomistic approach to warrant our using that label for our position."[37] Most importantly, it captures the deep

31. I have surveyed the biblical data in chapter 2, drawing heavily on Cooper. Chapters 3 and 4 considered the philosophical data.

32. Aristotle, *On the Soul*, 407b24 and *Metaphysics*, Book 8, Chapter 6.

33. Aristotle, *On the Soul*, 413a2–3.

34. Aquinas, *Commentary on Aristotle's De Anima*, II.1.225.

35. Koons, "Against Emergent Individualism," 378, 381–91. "Thomistic" identifies it with the insights of Thomas Aquinas.

36. For instance, see Moreland, "In Defense of Thomistic-Like Dualism," 102–22. A synonym Moreland sometimes uses is Thomistic-like organicism, as for example in Moreland, "Scientific Late Medieval Aristotelianism (Organicism)," 105–8.

37. Moreland and Rae, *Body and Soul*, 10. "Our" here referring to his and his co-author's view. Some are not happy with this interpretation of Aquinas. See Van Dyke, "Not Properly a Person, The Rational Soul and 'Thomistic Substance Dualism,'" 186–204. However, see Goetz and Taliaferro, *Brief History of the Soul*, 55 for a summary of Aquinas's view that "the soul is immortal and survives death" and how he integrates this with Aristotle's hylomorphism. This position can be supported by Aquinas's claims that "The human soul retains its own being after the dissolution of the body" (*Summa*

unity of soul and body, due to Aristotle and Aquinas's understanding of the relationship of form to matter.

As for the modifications of Aquinas's view to a "Thomistic-like" alternative, philosophers often do this to address deficiencies in the original thinker's account. Philosopher Karol Wojtyla (who subsequently became Pope John Paul II) did something similar as part of the school of thought known as *Lublin Thomism*. Wojtyla tried "to enrich [Aquinas's] one-sided aspect of man's rationality by including within the definition of human reality the entire range of human actions."[38] In this Wojtyla "accept[ed] the doctrine of St. Thomas on the composite of the soul and body of man"[39] and argued that doing so "leads to the acceptance of the existence of a substantial soul with its possibility of existing independently of the body."[40]

Taliaferro prefers the term *integrative dualism*,[41] stating, "According to substance (or integrative) dualism, a healthy, functioning human person lives and acts as a functional unity."[42] One more helpful term used for this view, by Millard Erickson in his systematic theology textbook, is *conditional unity*. He writes,

> The full range of the biblical data can best be accommodated by the view which we will term "conditional unity." . . . The normal state of a human is as a materialized unitary being . . . [which] can, however, be broken down, and at death it is, so that the immaterial aspect of the human lives on even as the material decomposes. At the resurrection, however, there will be a return to the material or bodily condition.[43]

All these terms—holistic dualism, Thomistic hylomorphism, Thomistic-like dualism, Lublin Thomism, integrative dualism, and conditional unity—capture the understanding of us as a deep functional unity and

---

Theologica I.76.I) and that the soul "has its own mode of existing superior to that of the body and not dependent upon it" (*Disputed Questions on the Soul*, 14).

38. Mieczylaw Krapiec, *I-Man: An Outline of Philosophical Anthropology*, 17.

39. Woznicki, *A Christian Humanism*, 17.

40. Woznicki, *A Christian Humanism*, 64, n. 14. For a very helpful summary of Lublin Thomism, see Cooper, *Body, Soul, and Life Everlasting*, 222–26.

41. "In all, I believe Aquinas's philosophy of mind is somewhere in the neighborhood of integrative dualism." Taliaferro, *Consciousness and the Mind of God*, 230.

42. Taliaferro "Substance Dualism: A Defense," 3, 43–60.

43. Erickson, *Christian Theology*, 555.

yet nevertheless an ontological duality.[44] Therefore, all are appropriate ways to refer to this anthropology.

Neurotheologians might raise a number of objections against this understanding of what we are. The next two chapters will consider three defenses of neurotheology (chapter 7) and three challenges to holistic dualism (chapter 8).

---

44. See Rickabaugh and Moreland, *Substance of Consciousness*, 314–30 for a fuller description of dualistic views. Willard suggests additional sources on anthropology, from both biblical and philosophical perspectives, in footnote 4 of *Renovation of the Heart*, 265–66. Finally, see Simpson, Koons, and Teh (eds.), *Neo-Aristotelian Perspectives on Contemporary Science*.

7

Three Common Defenses of Neurotheology

We must love souls in order to care for them.
—Dallas Willard[1]

Dallas changed my life with his teaching about loving souls. I returned the favor by telling him about attachment love in the human brain.
—Jim Wilder[2]

> **CHAPTER SUMMARY**
>
> Three common defenses of neurotheology are outlined and critiqued in this chapter. First, neurotheologians may appeal to the value and success of science in answering fundamental questions, including the question of what we are. This idea is known as "scientism," which is defined and shown to be inadequate for three reasons: it begs the question in favor of physicalism, it is self-defeating, and

1. Wilder, *Renovated*, 4.
2. Wilder, *Renovated*, 5.

science itself depends on knowledge discovered outside of science. Second, neurotheologians may argue that Dallas Willard, someone who thought deeply about what we are, became a neurotheologian later in life. This claim is refuted from Willard's writings throughout his career, and further clarification is drawn from conversations Willard had as he approached his passing with his friend, student, and protégé J. P. Moreland. A final objection is that neurotheology must have the right understanding of what a human person is, because it helps so many people. In response, I demonstrate that neurotheologians help others *in spite of* their view of what we are, rather than because of it, and that neurotheology is actually harmful to others if applied consistently.

IN CHAPTERS 5 AND 6, I outlined an understanding of the soul and its relationship to the body that I believe is the correct understanding of what we are, rather than the picture neurotheologians paint. However, I expect Thompson, Wilder, and others might raise six objections to my account: three in defense of neurotheology, and three in critique of the holistic dualism I endorse. These objections must be evaluated. In this chapter, I will consider the first set of objections—specifically, three defenses that may be offered on behalf of neurotheology.

DEFENSE 1: LET SCIENCE BE OUR GUIDE

First, neurotheologians may argue that neuroscience is providing more and more data on how our brains work. Therefore, they may contend, this should be at least the primary if not the only basis for developing an accurate anthropology.

This argument is often made by Nancey Murphy (whom we encountered in chapter 1), a professor of philosophy at Fuller Seminary. She states that her "approach to philosophical problems [such as the nature of the soul is] via science."[3]

Wilder follows this approach to understanding what we are (again, he did his graduate studies at Fuller). He tells us why he believes he must "modify" Willard's understanding throughout *Renovated*, perhaps most clearly when he says, "The neuroscience of attachment suggests some modifications of [Willard's] VIM model."[4] The VIM model (to which I will return in chapter 9) expresses Willard's core understanding of spiritual formation.[5] It begins with *Vision* (a desire to walk closely with God), followed by an *Intention* to become an apprentice of Jesus, which leads to embracing the *Means* of practicing spiritual disciplines. Yet Wilder replaces Willard's idea of "intention," claiming, "Intention becomes the wrong word for the *I* in VIM. I suggest 'impetus.'"[6] An intention is a mental event that occurs in the mind, whereas Wilder's impetus occurs in "the fast track in the right brain."[7] In other words, according to Wilder, neuroscience tells us that a brain state (an impetus) causes us to make a decision. Therefore, this scientific fact must replace Willard's understanding, based in philosophy, that decisions are guided by intentions of our soul.

Defining Scientism

I do not intend to disparage science in any way. Science is a wonderful source of knowledge. As mentioned in chapter 1, God has given us science as one way to learn truths about his creation. Those called to scientific fields have a high and noble calling, providing a great service through their scientific research, writing and teaching.

Yet this first objection takes science, which is *one* way of gaining knowledge, and absolutizes it as the *only* way to have knowledge. It implies that science alone gives us facts. On this basis, "discoveries" in other fields of study, such as theology and philosophy, are seen as only beliefs

---

3. Murphy, *In Search of the Soul*, 131.

4. Wilder, *Renovated*, 108.

5. Willard outlines the VIM model in his *Renovation of the Heart*, 82–91. He makes the point that this model applies not only to spiritual formation, but is "the general pattern that all effective efforts toward personal transformation . . . must follow." *Renovation of the Heart*, 82–83.

6. Wilder, *Renovated*, 127.

7. Wilder, *Renovated*, 148.

and opinions (unless they, in turn, can be justified by some form of "scientific" data). For those who are committed to this idea, whenever there is a disagreement between scientists and those in other fields of study, scientists must—by definition—be right. If this means rejecting contrary ideas discovered in fields such as biblical studies, theology, and philosophy, so be it! Even when these other fields seem to discover answers to a question that science cannot answer, scientism says, "No, that can't be right. *Eventually* we will find an adequate physical explanation!"[8]

This philosophical position concerning the nature of knowledge is known as *scientism*.[9] It is an epistemological[10] position concerning how we know what is true. Consistent with physicalism, if all that exists is physical, it follows[11] that we can verify something exists only if we can (at least in principle) observe it with one or more of our five senses. In other words, reality is only what we can confirm empirically. And since science is the study of that which can be empirically observed, we must look to science alone to answer our questions about what is real. From this assumption it follows that science *alone* gives us knowledge. As Wilfred

---

8. We see this, for instance, in the belief in a mythical "grandmother cell" as a physical placeholder until the actual physical means of storing memories in the brain is found. This is because, after years of neuroscientific studies, no correlation has been found between specific memories and specific neural events that occur each time a particular memory is recalled. One MIT neuroscientist, assuming that memories *must* reside in the brain, suggested that they are distributed across many regions of the brain, coining the term "grandmother cell" to signify this idea of a yet undiscovered physical storehouse of memories. See Gross, "Genealogy of the 'Grandmother Cell,'" 512–18.

9. There are actually two versions of scientism. Strong scientism maintains that science alone provides knowledge. Weak scientism grants that other fields also may provide knowledge, to a more limited degree, and only if its findings agree with those of science. Yet if the findings of science and those of any other discipline are in conflict, science must be right, and the findings of other discipline(s) must be modified accordingly. In either case, science is the final arbiter of what we do know and can know.

10. Epistemology is the sub-field of philosophy studying "the nature of knowledge and justification; specifically, the study of (a) the defining features, (b) the substantive conditions, and (c) the limits of knowledge and justification." Audi (ed.), *The Cambridge Dictionary of Philosophy*, "Epistemology," 233.

11. It can be debated which comes first, the metaphysic of physicalism or the epistemology of scientism. Lynne Rudder Baker observes that "Physicalism is the product of a claim about science together with a particular conception of science. The claim is that science is the exclusive arbiter of reality. . . . On this view, scientific knowledge is exhaustive." Baker, *Saving Belief*, 4. On the other hand, one may start with the assumption that reality is only physical, and from that derive an epistemology that limits what can be know to what can be known by scientific investigation. For my purposes here, I need not take a side in this debate.

Sellers famously quipped concerning this epistemology, "Science is the measure of all things, of what is that it is, and of what is not that it is not."[12]

From this approach to knowledge, it also follows that theology and philosophy are of no help in understanding what we are, or at least must be modified to fit the "scientific" picture of us as fundamentally physical. Therefore, we must reject all theological and philosophical arguments in favor of holistic dualism, because they are not "scientific." We must instead pin all our hopes on science to answer the question "What are we?"

To at least some degree, I believe the inroads of scientism among Christ-followers, along with our inherent desire for relevance and value, drive much of the interest in neurotheology among pastors, Christian counselors, and others. If in some way our ministries could become "scientific" by focusing on the brain, our culture would validate us as dealing with reality, rather than with outmoded matters that secular society has left behind. In this way, we could feel that our work is relevant and meaningful.

Three Reasons to Reject Scientism

But scientism is a failed epistemology. Much has been written to show its inadequacies. Due to space constraints, I can offer only a brief summary of three critiques.[13]

First, scientism is a classic case of the logical fallacy known as begging the question: assuming what you are trying to prove. The reasoning goes something like this: (1) all that is real, and therefore all that can be known, is physical; (2) science studies that which is physical; therefore (3) science is the only way to know what is real.

Of course, if one begins with the assumption that only physical things are real, then science *is* the only way to know truth. But why begin with this assumption? It originated with the Enlightenment, but there are no good reasons to believe that it is true.

---

12. Sellers, *Science, Perception and Reality*, 173.

13. See Moreland, *Scientism and Secularism: Learning to Respond to a Dangerous Ideology*. In his endorsement of this book, UCLA neuroscientist Jeffrey M. Schwartz writes that it "should be mandatory reading for serious Christians who want to intelligently engage in the interface of philosophy and science.... Moreland argues expertly that ... this central dogma of scientism erodes the serious pursuit of knowledge. Scientism isn't just poor science, it's poor thinking."

On the other hand, if one doesn't begin with this first assumption but allows for the possibility that immaterial things do exist, then one quickly realizes that these are not the kind of things that can be known by science. Rather, immaterial realities—including our immaterial dimension—will be known through other disciplines that study the immaterial realm, such as theology and philosophy. This alone is enough reason to reject scientism. Yet it faces even more serious problems.

Second, as mentioned in chapter 3, scientism is self-defeating: if it is true, it must be false. Scientism claims that science is the *only* way to know something. However, this knowledge claim itself must meet its own standard in order to be confirmed as true. In other words, the statement that "science is the only way to know something" must itself be proven scientifically. Yet this cannot be done. This idea cannot be observed in a beaker or tested through any other scientific apparatus. It is a philosophical belief about what can be known. So scientism fails its own test! Therefore, by its own standard, it must be judged to be false.

Only if one allows philosophical understandings to be true—at least this one about science alone giving us truth—could scientism be true. But in this case, the claim that "science alone gives us truth" is already falsified, for this one philosophical position is true. So either way, scientism collapses under the weight of its own condition for knowledge.

Third, science depends on a number of commitments that are not themselves scientific but are assumed to be true and support the entire scientific enterprise. Examples include the philosophical assumptions of the uniformity of cause and effect, that the future always reflects the past, that the laws of logic are valid, and that ethical values such as reporting data accurately should be employed. These are all fine assumptions. However, none of them are scientific assumptions. Rather, they are philosophical convictions that one must assume to be true before one can even get started doing science. Therefore, in the very practice of science, one must implicitly reject scientism. This is a third reason the epistemology of scientism can't be right.

Scientism must be rejected. Along with it goes Thompson and Wilder's approach to discovering what we are and how we flourish. As noted in chapter 1, they have looked only to scientists to develop their anthropology. As a result, they have made significant errors that led them astray. The solution is to also value the knowledge gained in other fields of study—in this case, theological and philosophical studies of what we are. If they had explored this rich history of discussion over the centuries, I believe they would have come to very different conclusions.[14]

Thompson and Wilder are not the only ones to make this error of discounting philosophical insights on the question of what we are. It is an epidemic among neuroscientists as well. As Bennett and Hacker observe,

> Cognitive neuroscience is sorely afflicted with various kinds of conceptual confusions. . . . Conceptual clarification is needed both for the identification and clear formulation of problems and for the description of the discoveries made and the realistic assessment of their significance. . . . What philosophy can contribute to neuroscience is conceptual clarification. . . . Philosophy can point out when the bounds of sense are transgressed. . . . It can make clear when the conceptual framework which informs a neuroscientist's research has been twisted or distorted.[15]

Doing so would have saved Wilder, Thompson, and everyone else who assumes scientism from many errors. "Following the science" is not a good defense of neurotheology.

DEFENSE 2: DALLAS WILLARD WAS A NEUROTHEOLOGIAN

Wilder's Interpretation of Willard

Jim Wilder is convinced that the ideas of Dallas Willard are consistent with neurotheology. Therefore, after each chapter in *Renovated* that contains one of Willard's talks at the 2012 Heart and Soul conference (which

---

14. Wilder grants that he "studied very little philosophy in school" (Wilder, *Renovated*, 31) and notes that he had hoped to study further with Willard, but didn't get the chance to do so before Willard passed away (Wilder, *Renovated*, 111). Yet this didn't lead him to suspend the writing of *Renovated* until he found someone else with philosophical training similar to Willard from whom he could learn. He does note that he read some philosophy books suggested by his brother, so as to avoid "total ignorance" (Wilder, *Renovated*, 31). Yet we are not told what these texts were, if he found them helpful, or even his brother's credentials that positioned him to recommend suitable texts, except that he reportedly "did his graduate work in philosophy and theology" Wilder, *Renovated*, 31–32). Furthermore, whatever books he read in philosophy and theology are not cited to support any of his arguments (whereas works by scientists such as Siegel are cited repeatedly). He seems to have concluded that science gives us the best information about what we are and how we flourish, and that therefore it was not important to thoroughly study or cite the findings of theology or philosophy when writing a book on this topic. All this is further indication of Wilder's scientism.

15. Bennett and Hacker, *Philosophical Foundations of Neuroscience*, 402, 405.

became the basis for Wilder's book), Wilder writes a chapter attempting to connect Dallas' insights to neurotheology, claiming that he is "adding, not subtracting [from]"[16] Willard's anthropology. Wilder justifies this undertaking by stating that "Dallas placed his endorsement on the first published copy of the Life Model" (Wilder's neurotheological approach to counseling).[17] If Willard endorsed neurotheology, this would indeed be a strong endorsement of its truth and value. But did he endorse neurotheology? Not in the least!

Why Wilder's Interpretation of Willard Is Deeply Misguided

A careful reading of *Renovated* reveals that Wilder is actually saying exactly the opposite of what Willard believes we are. Repeatedly, Willard states that *the soul* integrates all we are, and then immediately Wilder reinterprets this statement to mean that *the brain* integrates all we are. As mentioned above, Wilder states,

> Dallas describes the soul as 'that part of the person that integrates all the other dimensions to make one life.' The brain happens to contain a structure whose function is the integration of all internal states and all external connections with others.... When Dallas describes our experience of the soul ... he could hardly have described the cingulate [cortex] in clearer terms.[18]

In another passage, Wilder again conflates Willard's emphasis on the soul with his own emphasis on the brain: "The soul integrates our identities and directs the energy of everything it means to be human. The brain can create this integration using the cingulate cortex."[19]

Does Willard use "soul" and "brain" interchangeably as Wilder and other physicalists do? He does not, in *Renovated* or elsewhere. Quite to the contrary, he clearly and often states that human beings are souls that have bodies. He makes this point most forcefully in *Renovated* in a lengthy section[20] that describes the intellect, emotion, and will as aspects of the soul, which ultimately "puts all these parts together and makes

---

16. Wilder, *Renovated*, 149.
17. Wilder, *Renovated*, 108.
18. Wilder, *Renovated*, 85.
19. Wilder, *Renovated*, 89.
20. Wilder, *Renovated*, 53–63.

one life."[21] Willard concludes his discussion of the body by saying, "Your mind is not the same thing as your brain,"[22] thereby making it perfectly clear that he doesn't reduce the soul to the body.

Willard's dualism echoes through his other works on spiritual formation. For instance, in *Renovation of the Heart*, Willard emphatically states, "For all our fine advances in scientific knowledge . . . they tell us *nothing* about the inner life of the human being. . . . At most the sciences can indicate some fascinating and important correlations."[23] In *The Spirit of the Disciplines*, he observes,

> The dogmatic naturalist, sometimes under the guise of the latest "scientific thought," will insist that the human creature is only [matter]—nothing more, nothing less. . . . Yet, as creatures go, we *are* different. We *are* made for higher things. Our aspirations hint of such a truth. The age-old distinction between the body—the physique—and the person—the soul, spirit, mind—is rooted in the contrast between the unconscious physical facts of our lives . . . and our "conscious" life, our experiences, interests, meanings, thoughts, intents, and values.[24]

Willard was most certainly a holistic dualist, not a physicalist.[25] So what should we make of his endorsement of Wilder's neurotheological approach to counseling? It is physicalist in nature, yet Willard seems to think it is helpful. I have discussed this endorsement with J. P. Moreland, who is one of Willard's former students. Of their very close relationship, Moreland says, "Dallas and Jane adopted Hope and me as their family. He said I was one of his 'boys' (students about whom he cared and who were his main disciples). He was a mentor, friend and father to me from 1982

---

21. Wilder, *Renovated*, 57.
22. Wilder, *Renovated*, 63.
23. Willard, *Renovation of the Heart*, 17.
24. Willard, *Spirit of the Disciplines*, 46–47.
25. For more on Willard's holistic dualism, see his philosophical articles, especially related to Husserl, at www.dwillard.org, accessed January 21, 2024, as well as Willard, *Logic and the Objectivity of Knowledge*. Yet Willard's thinking on anthropology is scattered among a number of his writings, including some more popular books and articles in which he speaks loosely about anthropological matters, in order to encourage thinking about spiritual formation. For a summary, assessment, and nuancing of Willard's anthropology, see Moreland, "Tweaking Dallas Willard's Ontology of the Human Person," 197–202.

until his death."[26] In light of his deep and long friendship with Willard, Moreland states,

> I know for a fact that Dallas did not agree with Wilder's view due to its extreme and primary emphasis on the brain. He did, however, like the attempt to integrate brain studies with classic spiritual formation and I believe he is endorsing that effort. By saying Wilder's was the best we have, he is implying that we have done little work on this and Wilder's is a start. He would not endorse his view *tout court*. . . . He is being gracious and encouraging the effort.[27]

These comments are consistent with what Moreland says elsewhere. He writes that one of Willard's "two goals in formulating his anthropology [was that] he wanted his model to shed light on, allow for deeper insights into, and foster interest in spiritual formation, especially the role of the body in spiritual maturation."[28] For this reason, Willard valued any efforts to move away from the Gnostic tendency to deny the importance of the body. As we will see in the next chapter, this is why his understanding of spiritual formation is so insightful and helpful to so many.

Finally, Wilder seems to imply that he was able to help Willard understand and embrace neurotheology in the last few years of his earthly life. Wilder writes, "Dallas changed my life with his teaching about loving souls. I returned the favor by telling him about attachment love in the human brain."[29] This statement implies that not only did Wilder tell him about his physicalistic neurotheology, which I am certain he did, but that Willard in turn embraced it and believed it to be true, which I am certain he did not. In the last few months before Willard's passing, he invited Moreland to his home for one final visit. In their discussion they talked about Willard's anthropology. Of this conversation Moreland states, "Dallas was explicit that his view had not changed on the nature of a person."[30]

As mentioned earlier, Moreland delivered a moving eulogy at Willard's memorial service. As he recounted what was of utmost importance to Willard, he recalled, "[Dallas] had two main concerns. The first concern

---

26. J. P. Moreland, email message to author, November 10, 2023.
27. J. P. Moreland, email message to author, November 10, 2023.
28. Moreland, "Tweaking Dallas Willard's Ontology," 187–88.
29. Wilder, *Renovated*, 5.
30. J. P. Moreland, email message to author, November 10, 2023.

was that the spiritual formation movement be established on more intellectually rigorous philosophical and theological underpinnings."[31] Willard had given his life to defending the nature of persons as immaterial substances. This was the foundation of his work in spiritual formation, permeating each book he wrote on the subject. He did not change his mind as he approached his own death. Rather, Willard affirmed holistic dualism to the end of his earthly life (and I'm sure he does so even more now as he awaits the resurrection of his body!). This second attempt to bolster support for neurotheology fails as well.

DEFENSE 3: NEUROTHEOLOGY IS HELPING MANY PEOPLE

Many people have testified to being helped by neurotheologians such as Thompson and Wilder. Thompson writes, "I have noticed how much patients benefit when they have a better working understanding of the structures and function of the brain. It seems to help them see the connection between activity within their bodies and their thoughts, feelings, and behaviors. It also gives them a greater appreciation of what makes them uniquely human."[32] Wilder writes, "the brain's need for love-that-equals-attachment could explain why spiritual practices sometimes disappoint diligent Christians."[33] He then devotes his book's conclusion to illustrating how neurotheology has helped eleven people who developed a deeper understanding of how the brain works and thereby understood better how to grow in Christ and flourish.[34]

Neurotheologians Are Helpful in Spite of Their Neurotheology

But these people are being helped *in spite of* neurotheology, not because of it. Yes, our brains *are* actively involved in our flourishing, since we are a functional unity, as previously discussed. But the brain functions largely

---

31. Dallas Willard Ministries, "Dallas Willard Memorial Service, J. P. Moreland." Available at https://www.youtube.com/watch?v=AzSEeIUoksU&ab_channel=DallasWillardMinistries (from 4:00 to 4:36), accessed January 31, 2024.
32. Thompson, *Anatomy of the Soul*, 31.
33. Wilder, *Renovated*, 7.
34. Wilder, *Renovated*, 181–99.

in response to the soul's direction.[35] So as the soul matures, the brain increasingly functions in healthy ways.

However, with regard to the nature and practice of spiritual formation Thompson and Wilder get this fundamental order of the primacy of the soul wrong with their misplaced emphasis on the brain. Consistent with their physicalism, their emphasis is on *neural* formation rather than *spiritual* formation. Thompson says, "In short, the disciplines enable us to pay attention to our [brains]."[36]

Yet in spite of this error, they regularly slip into talking about *spiritual* formation, and they provide good advice about caring for our *souls* in order to properly use and modify our brains. Therefore, their advice *is* helpful to others, but *only as they are inconsistent with the physicalism of neurotheology* and focus instead on formation of the soul.

For example, consider the emphasis both Thompson and Wilder put on attachment—healthy relationships with God and others. Of such relationships, Thompson states, "There is nothing more crucial to our long-term welfare. In fact, virtually every action we humans take is part of the deeper attempt to connect with other humans. . . . Another terms that reflects this idea of connection is *attachment*."[37] Wilder adds concerning our relationship with God,

> Attachments are powerful and long-lasting. . . . Salvation through a new, loving attachment to God that changes our identities would be a very relational way to understand our salvation: We would be both saved and transformed through attachment love from, to, and with God . . . building attachment love is the central process for both spiritual and emotional maturity.[38]

However, our understanding of the brain adds nothing to our understanding of healthy relationships with others, including God. The ideas Thompson and Wilder express that are helpful in healthy relationships all refer to the soul, not the brain. For instance, over several pages Thompson

---

35. At times, the direction of causal influence and dependency also goes from brain to soul, as will be mentioned in chapter 9 concerning the value of pharmaceuticals. Yet the point here (and in chapter 9) is the irrelevance of understanding neuroscience to help in our *spiritual* formation.

36. Thompson, *Anatomy of the Soul*, 180. See chapters 3 and 4 for a discussion of Thompson and Wilder's equating and ultimate reduction of the mind (or soul) to the brain throughout their writings.

37. Thompson, *Anatomy of the Soul*, 109.

38. Wilder, *Renovated*, 6–7.

discusses research on attachment theory.[39] Yet the research he cites all has to do with children and adults developing and sustaining properties such as emotional stability, confidence, and security. Similarly, in summarizing the benefits of attachment love, Wilder speaks of sensing God's leading, paying attention to what is distressing his wife, sharing a sense of identity with others, regulating one's feelings, desires, impulses, and emotions, and responding to others in gentle, protective ways.[40] When they express such sentiments, Wilder and Thompson are discussing the various capacities of the soul, not the brain.[41]

So, when Thompson and Wilder talk about what our brain does, we learn nothing new about our spiritual formation,[42] but only about how the soul may be using the brain to accomplish its purposes. This information about the *brain* does not help us in shaping our *soul*. Instead, to be spiritually formed, we must understand the central role the *soul* plays in the process.

One example that Dallas Willard used in *Renovated*, briefly noted in chapter 6, illustrates well the central role of the soul. In his second talk at the Heart and Soul 2012 conference (which became chapter 4 of *Renovated*[43]), he describes how we, as souls exercising our will, use our body to engage the world and flourish. He states that we are "like someone using one of those big earth-moving machines. All they are doing is sitting there pushing a button or moving a little lever. They have integrated their will with the powers built into that machine. . . . The spiritual life works essentially in the same way."[44] Developing Willard's analogy further may

39. Thompson, *Anatomy of the Soul*, 113–16.

40. Wilder, *Renovated*, 110.

41. As discussed in chapter 3, brain states *correlate with* the various states of our soul, but are not *identical* to our soul. Rather, as discussed in chapters 4 and 6, we are ultimately a soul that *causes* these brain states.

42. Understanding our brains certainly helps in treating some mental disorders. As mentioned earlier, medications can help to deal with a number of problems people experience. Other neurological treatments involve the person in engaging his or her brain, such as neurofeedback (a type of biofeedback that measures brain waves to produce a signal that can help a person learn to self-regulate their brain functions and thereby treat conditions such as anxiety). Yet even in these cases, it is still *the soul* that uses the data to self-regulate in order to compensate for neural deficiencies, as discussed in chapter 7 and in various other sections of this book. My basic point is that a better understanding of our brains does not help the process of *spiritual* formation, contrary to Wilder and Thompson's claims.

43. Wilder, *Renovated*, 53–68.

44. Wilder, *Renovated*, 59–60. This passage should be read in the context of Willard's other discussions of what we are, cited throughout this book; it does not mean

help to clarify what he is saying. It will also clarify the neurotheologians' confusion, as well as why people are still helped in spite of this error.

Imagine that you buy land and hire a company to build a house. Plans are drawn up and construction begins. First, the ground must be dug up and the foundation poured. To do this, a large excavator is hauled to the site. It is a massive and very complex machine, capable of being maneuvered to precise locations and breaking through rock in order to excavate exactly the right amount of earth at the exact location determined by the builder. However, when the builder visits the worksite, she is shocked to discover that the excavation is off by ten feet, meaning that your house will be built partly on your neighbor's property! This will not lead to your or your neighbor's flourishing, to say the least!

She promptly informs the excavator operator of his error, and he immediately fills in the hole he dug, moves the excavator to the correct location, and begins digging the foundation there. As a result, the hole is dug precisely as desired, the foundation is poured, and the house is built.

Now suppose that, after you have settled in, you hold a housewarming party. There you recount this problem and how it was solved. An expert in mechanical engineering is present and begins explaining in great detail the mechanics of excavators, including how electrical impulses from the battery are linked to buttons and how the drive train connects to levers that control the way excavators are moved into precise locations to do their work. He says this is a special feature of excavators known as "geoplasticity"—i.e., their ability to change their locations in order to better accomplish their purposes. Of course, you know that it was the operator who initially had a wrong understanding of where to dig the foundation, was corrected, and then "pushed buttons and pulled levers" to move the excavator into the right position. Only the operator's choice to utilize the excavator's "geoplasticity" enabled the excavator to dig the foundation in the right place. Excavators are not autonomous robots—their every move is directed by a human operator.

But much to your dismay, the mechanical engineer scoffs at your explanation in light of the amazing advancements that have been made in mechanical engineering. No matter how many times you try to help him understand it was ultimately the operator who dug the foundation

---

that he endorsed the Cartesian view of the relationship between soul and body, a view that rejects the deep causal relationships between the two (to be discussed in chapter 8). The analogy he uses and that I expand here simply identifies the substantial soul as ultimate in this relationship and as using the body to accomplish its ends.

by using the excavator as a tool, he just keeps talking about how amazing the machine is. You finally give up, concluding he just doesn't have "eyes to see" the reality of the operator behind the operations of the excavator.

In the same way, Willard is arguing that you (a substantial soul) have this wonderful physical thing called a brain and body, in order to accomplish your will (and hopefully God's will) on this planet. This is the deep functional unity of persons discussed in chapter 6. Yet ultimately it is *you* who cause *your body*, including *your brain*, to engage the world in helpful or unhelpful ways by "pushing a button or moving a little lever" in your brain. You then use the feedback from your body and brain, just as the operator used feedback from the excavator, to know if you are being effective.[45]

Therefore, while certainly interesting, the brain's operations are not ultimately the root of the problem. So no matter how much we know about the mechanics of brains, what is ultimately necessary is people understand their wrong beliefs and choices they make as the "operators." Then they can correct these wrong beliefs and make proper choices, involving the brain in the process to accomplish their will. As Willard summarizes, "The part of us that drives and organizes our life is not the physical. This remains true even if we deny it. . . . The human spirit is an inescapable, fundamental aspect of every human being."[46] And yet there is a "positive role of the body in the process of redemption, as we choose those uses of our body that advance the spiritual life . . . we live only in the actions and dispositions of our body."[47]

We may now connect Willard's analogy to how neurotheologians help people, in spite of their anthropology. Thompson and Wilder's physicalism leads them, in the more theoretical parts of their books, to

---

45. The excavator analogy can be extended further. Suppose that the excavator was moved into position and begins digging, but then the hydraulic system beings to leak. This mechanical failure would cause the higher-order desire of the operator to accomplish his task to be thwarted by this lower-order blockage. The hydraulics must be fixed. But this alone will not enable the foundation to be completed; it will only allow the operator to resume doing what he wants to do. This is parallel to cases where the brain does not develop, so that people are incapable of doing what they want to do in the world (such as in cases of epilepsy). The brain must be restored to proper functioning, if possible. But this improvement only allows the person to live as he or she wishes (in the words of chapter 5, to "exemplify" his or her highest-order capacities). The person must still take action to do so. I will say much more about the role of the body in spiritual formation and human flourishing in chapter 9.

46. Willard, *Renovation of the Heart*, 13.

47. Willard, *Spirit of the Disciplines*, 40–41.

write much about the brain (the machine) rather than the operator (the soul) in spiritual formation. However, each time they shift to application, discussing how in fact we grow in Christ, they shift to talking about the operator: how to tend to the soul (without naming it as such) by encouraging their patients and readers to adopt true beliefs and make proper choices in order to flourish. They are giving the immaterial "operator"—the soul—correct counsel, though they are inaccurately saying that the cause of their help is understanding the "tool" the soul uses—the brain.

For instance, Thompson tells the story of a man named George and his daughter Kristin, with constant references to their beliefs and emotions (as indicated by the italicized words in the following passage):

> George was *worried* . . . addiction was not what [Kristin] *wanted* . . . she didn't *know how* to stop . . . we began to explore her *awareness* of what made her *anxious*. . . . I asked him to tell me about *his anxiety* . . . she began to *pay more attention* to what she was *sensing within* herself . . . what he was *experiencing within* himself . . . powerful forces *within his own mind*. . . . Like George, we can be *inattentive* to a great many things: our *thoughts* and *feelings*, the *nonverbal signals* we send and receive from others and ourselves, the *memories* from our developmental years. George was *oblivious* to the many things that were influencing his *experience* of God and life.[48]

Thompson helped them identify their wrong beliefs, choose to deal with these beliefs in appropriate ways, and find healing in their relationships with one another and with God. Throughout this process and his other examples, he focuses on helping patients develop that which is "inside" Kristin, George, and others. He is helping them form their souls in appropriate ways.

Of course, due to the deep causal relation between the soul and the body, this results in the soul reshaping the brain to accomplish this end (neuroplasticity). Unfortunately, this is where Thompson gets sidetracked. When he moves back to theory, he focuses on the results *in the brain* in his final analysis of the root problem and solution, rather than the *person using the brain* to flourish. From this he concludes that "ignoring these aspects of his brain's function resulted in his missing ways that God was attempting to capture his attention."[49] Returning to Willard's

---

48. Thompson, *Anatomy of the Soul*, 53–57; emphasis added to identify all the faculties of the soul Thompson describes in this example, despite his physicalism.

49. Thompson, *Anatomy of the Soul*, 57. See also Wilder's concluding chapter,

earthmover analogy, this is like saying that ignoring the mechanics of your excavator caused the foundation to be dug in the wrong place, and therefore the solution is to better understand your excavator's mechanical systems! No, the excavator engages the world properly, and we can use our brains to engage the world effectively, only when we choose to use these tools to engage the world to the proper end. If not, the problem is operator error, not the mechanics.

Thompson could have provided much of the same advice without ascribing the problem and solution ultimately to the brain. Using many of the same words and phrases he used above (italicized here), he could have said this:

> George, I won't start by prescribing you medication for your *anxiety*, because your problem is not ultimately neurological. Instead, you have *thoughts and feelings, powerful forces* which you are *experiencing within your mind* or soul that are the ultimate root of your anxiety. I want you to be *attentive* to these forces, which may be tied to *memories from your developmental years* that you may be *oblivious to* and which result in the *nonverbal signals you send and receive from others*, all of which *influence how you experience God and life*. By concentrating on changing your beliefs, your soul will, in turn, restructure your brain accordingly, since the brain can be reshaped in response to the soul's direction. Once we address this root problem, we may also want to consider medication as a supplemental treatment at that point. And Kristin, I know addiction *is not what you want*, but you *don't know how to stop*. And this is causing you great anxiety. Like your dad, though your brain is involved in what you are experiencing, your root problem is not neurological, but rather beliefs and experiences that make your soul *anxious*. I'd like for you to *begin paying more attention to what you are sensing within yourself*—in your soul—and for us to *begin exploring what you become aware of that makes you anxious*, so that you also can restructure your brain.

This summary of their problem and path toward recovery would have correctly and explicitly focused on the soul, and it would help Thompson's patients understand how this in turn relates to their brain's activity. Instead,

---

referenced above, for many more examples of how ultimately the formation of the soul, not of the brain, brings change in the people he assists, specifically by teaching them how to be in healthy relationships with God and others.

he depicts the problem as one of neurology. But he is helpful in spite of this, because he gives them good advice on how to care for their souls.

Wilder also helps others by focusing on *spiritual* formation, though he constantly attributes this process to understanding *brain* formation. He begins *Renovated* by stating that "attachment love is the central process for both *spiritual* and *emotional* maturity."[50] He goes on to say that this goal is accomplished by focusing on "*spiritual* exercises and human-*relationship* exercises . . . on building attachment *love* with God and with people."[51] These are all activities of the soul. Finally, after his discussion of the nature and importance of attachment love, when seeking to identify this love with brain states, in Appendix A[52] he suggests ways to develop this attachment love. But none of these is a brain exercise. They are again all *spiritual* practices of the soul!

In sum, Thompson and Wilder are certainly helping people. But this is only because they are helping *people*, not brain—operators, not excavators. Their advice is good, but only in spite of the physicalism of their neurotheology, not in virtue of it.

Why Neurotheology Is Ultimately Harmful

Although neurotheologians may be helpful in some immediate situations because they are inadvertently directing people toward caring for their souls, their ideas are harmful in the broader scope of things. They are working against precisely what Willard, in his final months, said must happen for the long-term health of the spiritual formation movement. He said that for people to receive lasting help, the spiritual formation movement must be established on more intellectually rigorous philosophical and theological underpinnings. In other words, only by developing a deeper understanding of the person from theology and philosophy could a movement encouraging deeper spiritual formation be sustained.

Yet our understanding of what we are is becoming thinner and thinner as the discussion of neuroscience dominates the discussion of spiritual formation. Brains are being given too much credit, and as a result we are not developing a better understanding of how to care for our *souls*.

---

50. Wilder, *Renovated*, 7 (emphasis added).
51. Wilder, *Renovated*, 7 (emphasis added).
52. Wilder, *Renovated*, 201–8.

Perhaps an even greater danger occurs due to Thompson and Wilder's constant and subtle vacillation noted above between passages in which they provide helpful counsel in the care of our soul and other passages in which they define us as physical beings. For instance, in chapter 3 I mentioned a passage in which Thompson criticizes atheists Pinker and Dennett as he makes the pastoral point that we can't reduce our desire for God to neurons.[53] Yet immediately after he criticizes Pinker and Dennett for their skepticism about the validity of religious experience, Thompson shifts from this pastoral context to explaining the *cause* of our desire to believe or not believe in God. There he states that our desires are *caused by* our brains.[54]

Likewise, as Wilder frames his book's focus, he states that his interest is "developing relational exercises to help people love each other. People with emotional wounds seemed particularly hampered in growing and sustaining loving, joyful relationships."[55] This pastoral passage captures an important truth. Yet in order to attain this important goal of sustaining joyful, loving relationships, he explains that we must understand the brain as the cause of healthy emotions and connections to others. He claims that "using brain science *behind* secure and joyful attachments"[56] is the key, focusing on "the brain functions that *determine* our character,"[57] which in turn is shaped by whom we love. Moreover, he adds later that whom we love is also a function of the brain: "*In the brain*, our social identities are at the core of character formation."[58]

These frequent oscillations between helpful advice in the care for our souls in Thompson and Wilder's *pastoral* passages and insistence that we are fundamentally brains in their *anthropological* passages makes

---

53. "One way to express their perspective is to say that if we can reduce our experience (in this case, of God) to that which we can measure (our genes and our neurons), we can eliminate the necessity of the God we thought existed." Thompson, *Anatomy*, 10.

54. "Most of us either want to believe in and have a relationship with God or we don't. Either way, we'll find ways for our left hemispheres to 'prove' what our right hemispheres are longing for—or are too terrified to desire." Thompson, *Anatomy*, 10.

55. Wilder, *Renovated*, 5.

56. Wilder, *Renovated*, 5 (italics added).

57. Wilder, *Renovated*, 6 (italics added).

58. Wilder, *Renovated*, 88 (italics added). He continues, "Brain scans reveal that the pain of rejection and abandonment is attachment pain down in the thalamus. Broken attachments upset the thalamus severely. Attack and withdrawal are signs of a brain in enemy mode. Enemy mode is the fast track [of the brain] operating without attachment. When the fast track has a broken attachment . . . it can become cruel . . . or even very dangerous."

Anatomy and *Renovated* very dangerous texts. Because their pastoral passages offer much helpful advice, it is easy for readers to also accept the physicalism endorsed in their anthropological passages. From this step, readers can be led to embrace the idea that a better understanding of neuroscience is the key to human flourishing.

As the examples mentioned in the introductory chapter illustrate, I have seen firsthand many Christian leaders, pastors, and counselors read these books and begin endorsing its underlying physicalist anthropology. In most cases, they don't seem to realize that they have adopted a faulty understanding of what we are. If asked directly, they often affirm that we are fundamentally immaterial beings. But in their sermons, counseling sessions, conference speaking, podcasts, and books they increasingly emphasize our brain functions as the core of what we are. Doing so only propagates this insidious anthropology among the many believers they in turn influence, leading to the harmful repercussions discussed in chapters 9 and 10.

With the failure of this third line of argument—that neurotheology is helping many people—we see that none of these three defenses of neurotheology hold up to scrutiny. Yet neurotheologians may reply that the alternative—holistic dualism—is an even more problematic understanding of what we are. The three most common objections to holistic dualism will be considered in the next chapter.

8

Three Common Objections to Holistic Dualism

Many of the popular objections against dualism are mistaken, unfair, or just plain bad philosophy.
—WILLIAM HASKER[1]

The time should be long past when scholars ... can get away with discrediting and dismissing dualism by simply equating it with Platonism.
—STEWART GOETZ[2]

> **CHAPTER SUMMARY**
>
> Neurotheologians may argue that even if physicalism has its problems, it avoids three more severe problems that holistic dualism entails. In this chapter I discuss and respond to each of these objections. First, neurotheologians may argue that we should always choose the simpler solution, for it is never helpful to make things more

---

1. Hasker, "On Behalf of Emergent Dualism," 95–96.
2. Goetz, "Is N. T. Wright Right about Substance Dualism?" 189.

> complex in our search for truth, and that physicalism is a simpler explanation of our nature because it posits that we are only one thing (a body) and not two (a body and soul). But this principle of simplicity is relevant and applicable only if we have no other reasons to prefer one view over the other. It is a tie-breaker if our investigation of competing views finds both to be equally defensible. But we have found good biblical and philosophical reasons to prefer holistic dualism over physicalism. Therefore, there is no tie to be broken, and so this objection fails. A second possible objection is that holistic dualism cannot explain how an immaterial and a material thing could interact with one another. Five responses are offered to show the failings of this objection. Finally, it may be objected that holistic dualism implies that animals also have souls, which could appear to be more in line with a pantheistic than with a Christian worldview. In reply, I discuss how the belief that animals have souls, understood as an immaterial dimension, is affirmed both in Scripture and throughout church history.

IN CHAPTER 7, WE saw that defenses of neurotheology fail. However, even though it has its problems, neurotheologians may still argue that it is better than the alternative! Accordingly, we also need to consider three often-raised objections to holistic dualism.

OBJECTION 1: KEEP IT SIMPLE

The first objection is the principle that we should always prefer simpler explanations to more complex ones. This is known as "Ockham's razor,"[3]

---

3. Ockham's razor, developed by philosopher William of Ockham (1287–1347), is also known as the Principle of Parsimony. For more on this objection and responses to it, see Goetz and Taliaferro, *A Brief History of the Soul*, 190–94; Moreland and Craig, *Philosophical Foundations for a Christian Worldview*, 244–45; Rickabaugh and Moreland, *Substance of Consciousness*, 298–99. Included in this last source is a fascinating discussion of the value of simplicity in finding truth in the first place.

for it focuses on cutting away the extra "fat" so that we are left with only the lean, and therefore the presumably true, understanding of reality.

Applied to the question of what we are, this argument would say that the physicalist's explanation of what we are is about as simple as it gets—we are just one thing (namely, matter or a physical body). The dualist adds two immaterial realities—consciousness and ultimately a soul—and thereby adds an unnecessary layer of complexity, a "ghost in the machine"[4] when the machine itself is all we need. Therefore, we should prefer the simpler explanation that we are fundamentally just a physical thing—a body. In the words of physicalist philosopher of mind Jaegwon Kim on this topic, "What can be done with fewer assumptions should not be done with more."[5]

This would certainly be a forceful objection if its underlying assumption—that the physical explanation of what we are is in fact adequate—were correct. However, as chapters 2, 3 and 4 showed, the physicalist explanation is far from sufficient, based on biblical and philosophical data. In the words of Goetz and Taliaferro, "We must not mistake simplicity for adequacy."[6] As we have seen in previous chapters, consciousness and a soul *must* be included to explain these data. Therefore, there is no excess "fat" in the dualist's understanding of our nature.

Furthermore, even if the physicalist's view is indeed simpler, simplicity is only one element that can make one position more reasonable than another. Other features of a view that may make it preferable include its accuracy, consistency, breadth of scope, and fruitfulness.[7] The dualist view gets high marks for a number of these other features (certainly accuracy, consistency, and breadth of scope) in light of all the data. Therefore, given this expanded list of desired features of a theory, dualism is again preferable. Ockham's razor fails to defeat holistic dualism.

OBJECTION 2: HOW CAN THE BODY AND SOUL INTERACT?

Perhaps the most common objection raised against any form of dualism is the *problem of interaction* (sometimes referred to as the *problem of*

---

4. Ryle, *The Concept of Mind*.
5. Kim, *The Philosophy of Mind*, 99.
6. Goetz and Taliaferro, *A Brief History of the Soul*, 192.
7. Kuhn, *The Essential Tension*, 321–22.

interactionism). Simply stated, this objection contends that the mind and the brain (or the soul and the body) are such different types of things that there is no way they could interact with one another. As physicalist Paul Churchland asks rhetorically, "How is this utterly insubstantial "thinking substance" to have any influence on ponderous matter? How can two such different things be in any sort of causal contact?"[8]

Five Replies to This Objection[9]

God's Interaction with His Creation

This objection runs into a number of problems. First, for the Christian this argument has little force, because we have independent reason to believe that God, an immaterial person, causally interacts with the physical realm (such as by causing the universe to begin and miracles to occur). Therefore, we know it is not impossible (or even improbable) that other immaterial things such as minds could also cause effects in physical things such as brains. On the other hand, if we *do* rule out the possibility of the mind causally interacting with the brain, we would also have to deny the possibility that God can causally interact with the world. These parallel cases of the immaterial interacting with the material stand or fall together.

Knowing That Without Knowing How

Second, again the distinction between knowing *that* and knowing *how* (discussed in chapter 5) comes into play. As another example, in quantum physics we know *that* quarks have certain causal effects, though we do not yet know *how* they do so. Similarly, we have seen that there are very good reasons to believe *that* the soul exists and is causally related to the body. Therefore, even if we cannot explain how this occurs, this does not discount our knowledge *that* in fact it does happen.[10]

---

8. Churchland, *Matter and Consciousness*, 9.

9. For a fuller treatment of interaction see Rickabaugh and Moreland, *Substance of Consciousness*, 276–83.

10. For more along these lines, see Bedau, "Cartesian Interactionism," 483–502. See also Foster, "In Defense of Dualism," 1–25.

Causation Is Sometimes Basic

Third, this objection presumes the need for an intervening mechanism that "connects" the mind and brain, and through which the mind "pulls the levers." But why assume that this is the case? It seems this is the type of causal relationship that is basic (immediate and direct).

We find other cases of direct causation in the world, so it is not unreasonable to think the causal relationship between mind and brain is of this basic type as well. Goetz and Taliaferro summarize, "Causation of any kind, even among physical objects, is ultimately mysterious . . . some causal powers will turn out to be basic, and not derived from more basic powers. If this is not a deeply vexing mystery for physical causation, it should not be one for nonphysical causation."[11] They are making the point that basic causation must ultimately underlie all causal chains. Otherwise there would be an infinite regress of causes, for each cause would in turn need an intervening mechanism, *ad infinitum*. In this case, there would be no causal chains at all, for the process would never be started. By analogy, this would be like trying to jump out of a bottomless pit—there would no way to "get a foothold" in order to start the process.

Yet causation does occur. So all causal chains must include, at the outset, a case of basic causation to get the chain of events started. Therefore, given how common basic causation is, we should not assume, as this objection does, that there must be an intervening mechanism in the case of causal relationships between the mind and the brain. Rather, these causal relationships between mind and brain are another example of basic causation.

Assuming Physicalism to Prove Physicalism

Fourth, to ask for some fundamental physical force by which the soul acts on the brain is to beg the question[12] in favor of physicalism. To understand how this is so, we must understand the four types of causes.

Let's return to the housewarming party described in chapter 7. As the conversation continues, you tell everyone how pleased you are with the

---

11. Goetz and Taliaferro, *A Brief History of the Soul*, 147.

12. As discussed in chapter 7, begging the question is a fallacy of logic in which the conclusion is assumed in one or more premises. Here it is the assumption that only physical things exist, which leads to the conclusion that the causal interaction between mind and brain *must* be physical.

craftsmanship of those who laid the flooring, installed the countertops, and mitered the crown molding. "That's what really makes this house!" you declare. At this point, your wife jumps in and says, "Yes, but they had beautiful wood and granite to work with. Without the lovely materials we chose, it wouldn't be what it is." Your architect adds, "That's all true, but ultimately it's the plans we drew up to specify how your dream house should be built that allowed the workers to do their magic with these materials in the first place." After some thought, your son pipes up, saying, "Yes, but Dad, before any of that we decided we wanted to live in this neighborhood. That's why we have this house!"

These are four very different explanations of your new house. Yet each person is right, for each is identifying one of the house's four causes. Your wife is identifying the *material cause* (the physical components). You are focusing on the expert craftsmanship of the workers as they worked with the materials (the *efficient cause*). Your architect, as you might expect, focuses on the immaterial structure that was made physical by the material and efficient causes; this is known as the *formal cause*. And your son rightly reminded you that the family's desire for a new home got the whole process started in the first place. This is the end or purpose for which the house was built—the *final cause*.[13]

The first two of these four causes (matter and energy) are material and therefore empirically verifiable. The second two, form/essence and end/teleology, are immaterial and therefore not observable through the five senses. Yet historically, all four causes were understood to be part of an adequate scientific explanation of everything we observe in the world. This expectation was based on the philosophy of science that underlay scientific inquiry from Aristotle until relatively recently.[14]

However, in 1620, philosopher Francis Bacon (1561–1626) suggested a new philosophy of science.[15] As part of the burgeoning secular age sweeping Europe in his day, physicalism and scientism were all the rage. And formal and final causes were not empirically verifiable. Therefore, he argued that they should not be part of a "scientific" explanation. Rather,

13. Aristotle first identified these four causes in his *Physics* II.3 and *Metaphysics* V.2.

14. For a summary, see Goetz and Taliaferro, *A Brief History of the Soul*, 28–29. For a discussion of the importance of this approach to science, see Lowe, *The Four-Category Ontology: A Metaphysical Foundation for Natural Science*; Ross, "The Fate of the Analysts: Aristotle's Revenge," 51–74. Bringing all we know into the practice of science is what Alvin Plantinga calls "Augustinian Science." See Plantinga, "Methodological Naturalism," 143–54.

15. Bacon, *The New Organon*.

science should focus exclusively on material and efficient causes.[16] Others agreed, and the new era of science began. The fact that now even most Christians simply assume that material and efficient causes are the only relevant explanations of physical reality is an indication of how steeped we are in Enlightenment philosophy, in this case the Enlightenment's philosophy of science.

Applied to causal relations in the soul and body, intervening mechanisms (material and efficient causes) certainly explain causation in the physical realm—the brain. Yet the "problem of interaction" assumes that there must *also* be intervening mechanisms to explain how the soul causes changes in the brain. This begs the question in favor of physicalism. Holistic dualists reject this physicalist assumption that demands an intervening mechanism—that requires material and efficient causes—to explain how the soul interacts with the body/brain. There is no intervening mechanism. Rather, the soul effects change in the brain through formal and final causality. The specific type of soul (the essential nature, which is the formal cause) naturally moves toward the exemplification of the soul's highest-order capacities at the first-order level (which is the final cause), as discussed in chapters 5 and 6.

Of course, the physicalist, already committed to the existence of only material and efficient causes, won't buy this. But the Christian knows that God himself acts in the world first and foremost as the final cause. Before he created matter and energy (and therefore before there was even the possibility of material and efficient causes), he caused the world to come into being due to his desire to create a universe that would display his beauty, love, and majesty, including the desire to create us as objects of his love. Therefore, the Christian has good reason to reject this physicalist assumption that only material and efficient causes are fully adequate explanations of what we observe around us, including causal relations between the mind and the brain.[17]

Interaction Is Expected in Holistic Dualism

Fifth, instead of interaction being a problem to be explained by holistic dualism, it is a fact to be anticipated. The problem of interaction has

---

16. For a nice overview of the shift to this new scientific method and critiques of it, see Losee, *A Historical Introduction to the Philosophy of Science*, 60–69.

17. Goetz and Taliaferro, *A Brief History of the Soul*, 156–76.

historically been raised against another form of dualism already mentioned briefly: Cartesian dualism. This is the version promoted by Rene Descartes (1596–1650), hence the name "Cartesian,"[18] though actually it has its roots in Plato's radical separation of the material and immaterial realms[19]—which also helped give rise to Gnosticism, discussed in the introduction to this book.[20] Briefly put, for Cartesian dualists, we are a soul that is "in" our body much as coffee is in a cup. The soul and body, like the coffee and the cup, are only superficially related or "united" to one another. The cup is what it is whether or not the coffee is in it. Similarly, the body is what it is regardless of the presence or absence of a soul. The body is its own substance, functioning on its own steam, and not as a result of being ensouled. Yet the body serves temporarily as the soul's "container," housing this second, immaterial substance for a time.

Therefore, due to the superficial relationship between soul and body, interaction between the two is surprising. It is natural to ask for some explanation of this interaction. Providing an adequate explanation to solve the "problem of interaction" has always been a challenge for Cartesian dualists.[21]

However, holistic dualism does not start with two very different things—soul and body. Rather, it starts with a soul that naturally creates and constantly interacts with its body. As Willard says, the advocate for holistic dualism is "denying that the body is 'just physical,' just some more or less mechanical device incidentally associated with a purely spiritual mind or self."[22] Rather, from the beginning the soul and body are deeply intertwined.

As an analogy of this deep unity and causal connections, think of brewing a cup of coffee. When the coffee grounds are placed in hot water,

---

18. See Descartes, *Meditations*, "Sixth Meditation: Of the Existence of Material Things, and of the Real Distinction between the Soul and the Body of Man," 150–69.

19. For how this applies to Plato's understanding of the soul and body, see especially his *Phaedo* dialogue.

20. For a summary of Plato's overall understanding of reality, see Copleston, *A History of Philosophy*, "Plato," 127–265. For the role that Plato's thought played in Gnosticism, see Latourette, *A History of Christianity: Beginnings to 1500*, 123–25.

21. Descartes proposed locating the connection in the pineal gland, consistent with his belief that a specific location was needed for material and efficient causation to occur. See Descartes, "The Passions of the Soul," 343–44. Some thoughtful Cartesian dualists have sought to resolve this problem in other ways. Notable among them are Richard Swinburne, Keith Yandell, and John Foster. See Swinburne, "Cartesian Substance Dualism," 133–51; Yandell, "A Defense of Dualism," 551–53; Foster, *The Immaterial Self*.

22. Willard, *Spirit of the Disciplines*, 82.

the water causes the grounds to partially dissolve, infusing the hot water with its glorious coffee aroma and taste. The water causes the grounds to partially dissolve, and the grounds cause the water to have a new smell and taste. Though imperfect, this analogy illustrates how the causal relationship between soul and body is much deeper than the superficial "coffee in a cup" unity of Cartesian dualism.

As a result of this deep unity, interaction is to be expected. In fact, it would be a problem for holistic dualism if there were no interaction.[23] This is precisely why Cooper chose the term holistic dualism, "to capture . . . the unity of human nature."[24] So, in the case of holistic dualism, the "problem" of interaction either evaporates (because it is expected) or at least is explained by the formal and final causes at work in the relationship between soul and body.[25]

Three Reasons Why This Objection Persists

Summarizing the problem of interaction, Hasker observes, "Many of the popular objections against dualism are mistaken, unfair, or just plain bad philosophy. For instance, the well-worn objection that mind and matter cannot interact because they are different kinds of substances has my vote for being the most overrated philosophical objection of all time."[26] So why is the "problem of interaction" so often trotted out and rehashed? And why is holistic dualism, along with its answer to the problem of interaction, not included in this conversation? I think there are three reasons why holistic dualism is almost always ignored, with the result that the problem of interaction is offered as a strong reason to discount dualism.

---

23. Moreland, "A Defense of a Thomistic-Like Dualism," 102–31.

24. Cooper, *Body, Soul, and Life Everlasting*, xxvii.

25. I have not considered here a newer and less frequently argued version known as emergent dualism (see for instance Hasker, "The Case for Emergent Dualism," 62–73), in which the soul itself emerges from a complex brain (rather than only properties emerging, which is the view of nonreductive physicalism). Though emergent dualism is similar to holistic dualism in also positing a substantial soul that is deeply united with the body, the physical origin and sustenance of the soul leave it susceptible to many of the same criticisms that are fatal to nonreductive physicalism (see chapters 2, 3 and 4). Therefore, I believe it fails as an adequate alternative.

26. Hasker, "On Behalf of Emergent Dualism," 95–96.

Ignorance

The first reason is ignorance. Many are simply not familiar with holistic dualism. Therefore, they assume Cartesian dualism is the only dualistic option. And since interaction is a significant problem for Cartesian dualism, the problem of interaction gets much press.

This is understandable. Many psychologists, psychiatrists, and other mental health practitioners have almost no training in the philosophical underpinnings of anthropology. A UCLA neuroscientist did a study of nearly 80,000 mental health professionals from around the world and found that a mere 2% to 5% had heard even just one presentation on the philosophical issues related to the mind or soul.[27] No wonder so many assume that Cartesian dualism is the only game in town and therefore dismiss dualism based on the problem of interaction.

Thompson and Wilder seem to fall into this category. Thompson refers to no holistic dualists in his discussion of what human beings are or, as far as I can tell, in his bibliography. If he was aware of holistic dualism, it seems reasonable to assume that he would have at least mentioned it, as this view makes a great deal of sense of the neuroscientific data he cites (as discussed in chapter 6). And even though Willard explicitly outlines holistic dualism in his chapters of *Renovated*, Wilder clearly misses this main point. Ignorance is the most charitable explanation of Thompson and Wilder's failure to engage holistic dualism.

Misunderstanding

There are others who I believe must have come across holistic dualism. Yet they seem to have misunderstood Aristotle's ideas concerning the relationship of form and matter, on which holistic dualism is based and runs through the likes of Thomas Aquinas, C. S. Lewis, Dallas Willard, and J. P. Moreland. Therefore, due to this misunderstanding they falsely identify all Greek thought with Plato's idea that matter and spirit are very separate, which they reject. From this they also reject what they understand to be the only anthropology arising from Greek though: Platonic/Cartesian dualism.

For instance, we see this misunderstanding in the writings of the usually very insightful N. T. Wright. He equates anthropological dualism

---

27. Siegel, *Mindsight*, 50–51.

with "a radical twofoldness of soul and body,"[28] echoing Plato's radical distinction between matter and spirit. He then claims the biblical account of what we are "has nothing whatever to do with Platonic or quasi-Platonic dualism," going on to say,

> An anthropological dualism tends to devalue or downgrade the body [and therefore] Christians should . . . resist attempts to reinstate a . . . dualism in which 'mind' becomes the significant reality rather than 'body' . . . [for] the implicit devaluation of the body and over-evaluation of the mind [as a result of dualism] has been a major problem in the western world for many generations and I would hate to think of this being simply pushed further.[29]

These are implications of Plato's ideas, but not of Aristotle's thought. Yet Wright's misunderstanding of dualism to mean only Platonic/Cartesian dualism leads him to dismiss dualism *per se*.[30]

In his summary of Wright's misunderstanding Goetz states, "The time should be long past when scholars like Wright can get away with discrediting and dismissing dualism by simply equating it with Platonism."[31] Theologians, pastors, and all thoughtful Christians must correct this misunderstanding by developing a better grasp of the ideas Aristotle brought to the anthropological table and have been developed by Aquinas, Willard, Moreland, and others.

---

28. Wright, "Mind, Spirit, Soul and Body."

29. Wright, "Mind, Spirit, Soul and Body."

30. Although elsewhere Wright affirms we live on after our deaths in a disembodied intermediate state, which can only be true if we are an ontological duality. He writes, "All the Christians departed are in substantially the same state, that of restful happiness. Though this is sometimes described as sleep, we shouldn't take this to mean that it is a state of unconsciousness [for] Paul . . . described life immediately after death as 'being with Christ, which is far better.' Rather, *sleep* here means that the *body* is 'asleep' in the sense of being 'dead,' while the real person—however we want to describe him or her—continues." Wright, *Surprised by Hope*, 171. Of such inconsistencies Cooper observes, "There are those who do not seem to sense the tension. I have heard a number of preachers and professors who in the very same sermons and lectures both denounce the body-soul distinction as an unbiblical Greek idea and nevertheless affirm that believers fellowship with the Lord between physical death and the final resurrection. They seem blissfully unaware that they are deploying contradictory assertions." Cooper, *Body, Soul, and Life Everlasting*, 4.

31. Goetz, "Is N. T. Wright Right about Substance Dualism?" 189. See also Rickabaugh, "Responding to N. T. Wright's Rejection of the Soul," 201–20.

Timidity

A third reason why holistic dualism is not considered seriously is sociological in nature. As we now live in a secular age, we tend to share the starting assumption that any talk of immaterial realities deeply united with what is physical is no longer fashionable. It is passé. It is outdated. It is ... so medieval! So away with Aristotle! Away with Aquinas! Away with any ideas based on their thinking, including holistic dualism!

But strip away the underlying philosophical assumptions (the ontological assumption of physicalism and the epistemological assumption of scientism) and the force of this sociological objection evaporates. Sadly, dropping these assumptions is unthinkable for many in the modern age. As atheist philosopher Daniel Dennett remarks, "dualism is to be avoided *at all costs*."[32]

However, truth is not a popularity contest. No matter what derogatory names may be thrown at an out-of-vogue theory, truth wins. This is why C. S. Lewis, who based much of his thinking on Aristotle and Aquinas, happily described himself as a believer in old and socially discarded yet still very true ideas.[33] Believing only what is currently in style is what Lewis rightly called "'chronological snobbery,' the uncritical acceptance of the intellectual climate common to our own age and the assumption that whatever has gone out of date is on that account discredited."[34] I am hopeful that many Christians today will reject chronological snobbery and instead, as Lewis did, swim against the cultural currents in pursuing truth and not fashion, truly "thinking Christianly" despite the swelling tide of secularism.

A Case of the False Dichotomy Fallacy

Regardless of the reason for disregarding holistic dualism (ignorance, misunderstanding, or timidity), to do so is to commit the fallacy of the false dichotomy. As mentioned earlier, this is an error of reasoning in

---

32. Dennett, *Consciousness Explained*, 37 (emphasis in original).

33. Lewis was a professor of medieval and Renaissance literature. A witty defense of the "old" and yet true is found in his inaugural lecture at Cambridge University, "De Descriptione Temporum." The lecture is available online at https://files.romanroads-static.com/old-western-culture-extras/DeDescriptioneTemporum-CS-Lewis.pdf, accessed November 20, 2023. See especially page 7.

34. Lewis, *Surprised by Joy*, 207–8. See also Lewis, *God in the Dock*, 200–207.

which two views are offered, one is shown to be false, and so the other is declared the winner. However, if there is another (or several other) options beyond the two offered, the reasoning that one of the two must be true collapses.

This is exactly what is happening here. Two fundamental views of what we are take the stage. On one hand is Cartesian dualism, which says we are composed of radically discontinuous material and immaterial dimensions. On the other hand is some form of physicalism.[35] Cartesian dualism is thought to be proven false due to the problem of interaction. Therefore, the alternative—physicalism—must be true.

Yet as we have seen, there is a third viable option between these two: holistic dualism. This view finds the "middle way," affirming both our functional unity *and* our ontological duality. Therefore, even if the problem of interaction defeats Cartesian dualism, there remains another option besides physicalism. Saying that the problem of interaction is a reason to reject all forms of dualism is an evident illustration of the false dichotomy fallacy.[36]

OBJECTION 3: WHAT ABOUT ANIMAL SOULS?

A final objection often raised against any view based on Aristotle's hylomorphism, including holistic dualism, is that it seems to imply that animals also have souls. Yet some may object that this sounds more like a pantheistic idea, consistent with Hinduism and perhaps Buddhism, but not a Christian view of the world.

---

35. As discussed in chapters 3 and 4, there are different forms of physicalism. Reductive physicalism holds that we are *only* one physical thing, while nonreductive physicalism claims that we are *fundamentally* a physical thing that also emits immaterial properties like thoughts. There are also other physicalist anthropologies such as the "constitution" view held by Calvin University philosopher Kevin Corcoran, "The Constitution View of Persons," 153–76. I have not explored this variation of physicalism, for (1) it does not seem to be the view of Thompson and Wilder, and (2) it shipwrecks on the same sandbar as other forms of physicalism which I have discussed. For specific critiques of constitutionalism, see Goetz, "A Substance Dualist Response," 177–80; Moreland and Rae, *Body and Soul*, 192–96.

36. It may also be guilty of the straw man fallacy, which involves stating a weak form of someone else's position that one can easily show to be false. Cartesian dualism seems like an easier target than holistic dualism, at least in terms of interaction, and so perhaps that is why so many treat it as synonymous with dualism. I don't know the motives of those who ignore holistic dualism, so I cannot say definitively whether they are knowingly engaging in the straw man fallacy.

First, the biblical record indicates that animals, as well as humans, do in fact have souls. In chapter 2, we explored Genesis 2:7, which says that God, in his creation of Adam, "breathed into his nostrils the breath of life; and man became a living being." The word used for "being," and in other translations as "soul," is *nephesh*. Yet the same word is used of animals in Genesis 6:17, Genesis 7:22, and Ecclesiastes 3:19. In the New Testament, the Greek word used repeatedly for soul (*psychē*) is also used in Revelation 8:9 of animals.

Of course, this doesn't mean animals have souls that are like ours. Specifically, they do not share the image of God, which we possess uniquely and which gives us special status among all of God's creation (as discussed in chapter 5). But from this, it does not follow that animals have no type of individuated nature that includes capacities to at least think, feel, desire, and interact socially.

This is what I observe in my dog. I often watch her remembering where she left her ball and running to get it. I see her full of joy when I come home. It is obvious that she desires to please me by obeying (though sometimes other desires overrule this one!). She is excited to see the dogs who live down the street and engages them in play, a social activity. She is clearly more than a machine. She is ensouled. Lewis has similar intuitions: "It is certainly difficult to suppose that the apes, the elephant, and the higher domesticated animals, have not, in some degree, a self or soul which connects experiences and gives rise to rudimentary individuality."[37]

In fact, until physicalism took hold in the seventeenth century, believers always assumed that animals could not be reduced to machines but also had an immaterial dimension—a "soul" of sorts. For instance, Augustine argued that all living things have a soul.[38] As Moreland and Habermas summarize,

> It wasn't until the advent of seventeenth-century Enlightenment . . . that the existence of animal souls was even questioned in Western civilization. Throughout the history of the church, the classic understanding of living things has included the doctrine that animals, as well as humans, have souls.[39]

---

37. Lewis, *The Problem of Pain*, 121. Aristotle's discussion of types of souls appears in his *On the Soul*, 413a23 to 435b26. For a summary of Aristotle's understanding of types of souls, see Willard, *Spirit of the Disciplines*, 57–58.

38. Augustine, *The City of God*, VII.29; Augustine, *The Trinity*, X.4.6.

39. Habermas and Moreland, *Beyond Death*, 293.

The denial of animal souls comes, in large measure, from Descartes's understanding of bodies as machines, not the holistic dualist view.

In conclusion, these three common objections to dualism fail.[40] Rather than objections burying substance dualism, it is actually gaining momentum in academic circles. In the opening chapter of their meticulously researched *The Substance of Consciousness*, Moreland and Rickabaugh state, "It is no exaggeration to say that substance dualism is undergoing an unforeseen revival and is poised to make a strong return in the 21st century [due to] the rapid growth in sophisticated new work on substance dualism."[41]

Therefore, in light of the well-grounded theological and philosophical data that we have discussed in some detail, we can be confident that each of us is an immaterial person deeply united to our body. It remains to apply holistic dualism to questions of human flourishing. I will explore some of these implications in the final two chapters.

---

40. For a discussion of other arguments raised against substance dualism, see Rickabaugh and Moreland, *Substance of Consciousness*, 275–307. For a detailed critique of physicalism and support of substance dualism in general, see Taliaferro, *Consciousness and the Mind of God*.

41. Rickabaugh and Moreland, *Substance of Consciousness*, 5.

9

Soul, Body, and Loving God

The spiritual and the bodily are by no means opposed in human life—they are complementary.
—Dallas Willard[1]

What you allow to occupy your mind will sooner or later determine your feelings, your speech, and your actions.
—Daniel G. Amen, M.D. and Lisa C. Routh, M.D.[2]

> **CHAPTER SUMMARY**
>
> We now move to practical application of the issues discussed in this book. Chapter 9 applies a proper understanding of ourselves to how we can best love God; chapter 10 will focus on how we can best love others. In this chapter, I first review the two ends of the spectrum, physicalism and Platonic/Cartesian dualism, to show how holistic dualism serves as a responsible middle way of understanding how we flourish. I apply the idea

---

1. Willard, *Spirit of the Disciplines*, 75.
2. Amen and Routh, *Healing Anxiety and Depression*, 184.

> that our soul has a hierarchy of capacities to our growth
> in Christ, and to how the spiritual disciplines relate
> to our flourishing. I also discuss how understanding
> the causal relationships among our soul's faculties
> can assist our spiritual formation. I conclude with
> some thoughts on the benefits of understanding
> the natural goal toward which our souls strive.

MARTIN LUTHER ONCE OBSERVED that our fallen nature causes us to veer off to extremes as we seek to live out our faith. In his words, it is like a drunk man trying to ride a horse—he is always falling off to one side or the other.[3] One such extreme is the tendency to reduce or fully reject the value of the physical realm, including our bodies. (This is the Gnostic error, based on Plato's radical bifurcation of matter and spirit, that we discussed in the introduction, and it undergirds the Cartesian dualistic view of what we are, as discussed in chapter 8.) This form of reductionism has led to grave errors, such as extreme denial of bodily needs, devaluing our daily work, and failing to care for God's creation.

In seeking to avoid these errors, many have appropriately emphasized the value of the physical realm in general and of our bodies in particular. However, it is also easy to go too far in this direction, falling off the other side of the figurative horse into implicit or even explicit physicalism and denying the reality or value of our immaterial self. This is what the neurotheologians have done. This physicalist reductionism leads to equally grave errors, including providing no motivation or strategies for *spiritual* formation.

The corrective to both these reductionistic understandings of what we are and their implications is to maintain a middle way between the two extremes. Holistic dualism is that middle way. It does not reduce us to either a soul or a body but affirms both as valuable dimensions of what we are and how we flourish.

Up to this point, we have seen why believers should embrace holistic dualism on biblical and philosophical grounds. Now it is time to consider

---

3. Luther, *The Table Talk of Martin Luther*, 631.

some practical implications of holistic dualism for human flourishing. In these last two chapters, I will suggest ways in which understanding holistic dualism helps us fulfill Jesus' greatest commandment as presented in Matthew 22:37-38: to love God (chapter 9) and our neighbors (chapter 10).

The examples I offer in these last two chapters are only a few of the many ways in which we can "think Christianly" about loving God and others in light of holistic dualism. I hope you will be stimulated to identify additional applications relevant to your particular context.

THE PLATONIC/CARTESIAN DUALIST REDUCTION TO "PURE" SPIRITUALITY

The Platonic/Cartesian reduction has led many throughout history and today to deny that the body has anything to do with our growth in Christ. If the soul is only superficially related to the body, our spiritual formation is purely an activity of the soul. Therefore, we grow in Christ only through "spiritual" activities such as being involved in church, prayer, and Bible study. What we do or don't do with our bodies makes no difference at all. Therefore, "mundane" activities—such as being an accountant, reading a Tom Clancy novel, or enjoying a backyard barbeque with friends—are not considered part of our spiritual lives (unless they can somehow be connected to evangelism or discipleship).

But the body *is* important to our growth in Christ. For instance, Job said that he had made a covenant with his eyes (Job 31) as part of loving God. The psalmist speaks of his body longing for God (Ps 63:1), Paul exhorts us to present to God our body as "a living and holy sacrifice" (Rom 12:1) and to "glorify God with your body" (1 Cor 6:20). The world around us is valuable and so God commanded us to care for it (Gen 1:26-28) and to enjoy the life God has given us on this earth (as Ecclesiastes 8:15 concludes: "So I commend the enjoyment of life"[4]).

THE PHYSICALIST REDUCTION TO SHAPING OUR BRAINS

On the other hand (or, to use Luther's analogy, on the other side of the horse) are those who disregard the soul. For people coming from this perspective, the idea that spiritual formation as essentially soul formation

---

4. NIV translation.

must be rethought, because we are ultimately physical beings. This is the error of the neurotheologians, who divert our attention away from spiritual formation and to brain formation, as they reduce beliefs, desires, choices, and other capacities of the soul to mere neural events.

As will be illustrated throughout this chapter, the implications of this error are equally, or perhaps even more devastating to our spiritual lives than those of the Platonic/Cartesian error, for as we have seen we *do* have a soul. Therefore, denying its existence makes it increasingly difficult to pursue or maintain a robust spiritual life, for the aim of spiritual formation is to form our souls—*spiritual* formation—into conformity with the image of Christ (Rom 8:29). Only by shaping our souls, in relation to our bodies, will we enjoy intimacy with God and exemplify the fruit of the Spirit, as our soul's highest-order capacities are fully exemplified in the lives we live.[5]

THE MIDDLE WAY: REDUCING TO NEITHER BODY OR SOUL

In Romans 6, Paul discusses how we use our bodies to live in the world, either glorifying God or dishonoring Him. Paul exhorts us to "not let sin reign in your mortal body so that you obey its lusts, and do not go on presenting the members of your body to sin as instruments of unrighteousness; but present yourselves to God as those alive from the dead, and your members as instruments of righteousness to God" (Rom 6:12–13). In 1 Corinthians 6:13 and 15, he reiterates that "the body . . . is for the Lord, and the Lord for the body . . . our bodies are parts of Christ."

For instance, we can use our eyes to look at that which is edifying to us and honoring to God, or to look at things that are not edifying or God-honoring. But it is not our eyes that are holy or sinful in the looking; rather, *we* use our eyes to honor or dishonor God. The same can be said of all other body parts. I (my self or soul) use my bodily parts, including my brain, to act. This is why my body is vitally important to my spiritual formation (or deformation).

Dallas Willard has influenced many to recover the importance of our bodies in spiritual formation, while still valuing the role of our

---

5. Willard suggests McNeill, *A History of the Cure of Souls*, for "a thorough and careful study of what we today have forfeited by our neglect of the soul." Willard, *Renovation of the Heart*, 266, note 7.

souls—a healthy "both/and"—due to his holistic dualism.[6] He has emphasized how activities that appear primarily spiritual in nature, such as Scripture memory and prayer, are obviously important. But so are primarily physical activities such as fasting and service. In fact, in all spiritual disciplines, both body and soul are involved to some degree.[7] As Willard puts it, "Every spiritual discipline, even things like memorizing Scripture, involves physical energy or power. You engage your body in it."[8] Praying on our knees helps us embody an attitude of humility and submission before God. Fasting helps us develop self-control to conquer lust. The way in which a sanctuary engages all our senses has a positive influence on worship. Love of God is increased by the things we see, the music we hear, the words we say, the people we are with, and so on.

These examples illustrate the two-way causal relationships between body and soul in our spiritual formation, all grounded in the anthropology of holistic dualism. God created us as embodied souls, and we love God most integrally and authentically when body and soul are united in worship. As Willard summarizes, "The physical human frame as created was designed for interaction with the spiritual realm."[9] Again, "It is the nature of the human being that the 'inner reality of the self' settles into our body, from which that inner reality then operates in *practice*."[10]

In fact, as discussed in chapter 2, the unity of our body and soul is so important in expressing our love for God that, after the death of our body and our temporary existence in an intermediate, disembodied state, God will reunite our souls and (resurrected) bodies to live and worship him forever in the new heaven and new earth (Rev 21). This is God's final stamp of approval on the fact that our bodies are crucial to our spiritual life.

---

6. Willard's influential writings on spiritual formation include *The Spirit of the Disciplines*, *Renovation of the Heart*, and *The Divine Conspiracy*. He also helped found Renovaré, a ministry designed to help believers "become more like Jesus" (see www.renovare.org, accessed January 31, 2024).

7. A spiritual discipline is anything we discipline ourselves to do repeatedly in order to grow spiritually. Willard's list includes disciplines of abstinence (solitude, silence, fasting, frugality, self-restraint, secrecy, sacrifice) and those of engagement (study, worship, celebration, service, prayer, fellowship, confession, submission). Willard, *Spirit of the Disciplines*, 158. For another seminal discussion of spiritual disciplines, see Foster, *Celebration of Discipline*. Of course, God is also at work in us to drive this process of spiritual development. As Paul said, "Continue *to work out* your salvation with fear and trembling. For *it is God who works in you*" (Phil 2:12 NIV, emphasis added).

8. Wilder, *Renovated*, 64.

9. Willard, *Spirit of the Disciplines*, 77.

10. Willard, *Renovation of the Heart*, 165–66.

One of my main underlying concerns in this book is to encourage believers to develop a more and more sophisticated understanding of what we are, absolutely essential for our spiritual formation. Here is one passage in which Dallas Willard emphasizes this point in some detail:

> *The nature of the person* is, today, a battlefield of conflicting academic, scientific, artistic, religious, legal, and political viewpoints.... We must understand that in today's "Western culture" the "academic" is never "merely." It is the academic that today governs the idea systems of our world and opposes traditional views of human nature—specifically the Judeo-Christian or biblical understanding of human life. Today you will hear many presumably learned people say that there is no such thing as human nature . . . [but] the issue of human nature is of great importance—too important for us to leave alone. We must deal with it if we are going to have anything useful to say about spiritual formation or the spiritual life that Jesus brings. Otherwise what we say will have no relation to the concrete existence of real human beings. . . . Once we clearly acknowledge the soul, we can learn to hear its cries.[11]

In chapters 5 and 6 I summarized the contours of a more nuanced understanding of the soul and body. I will now briefly apply this holistic dualism to growth in Christ.

Spiritual Formation and Our Capacities

In chapter 5, we noted that the soul possesses capacities that are arranged hierarchically. This helps us understand how we become conformed to the image of Christ and thereby flourish. It is by exhibiting all these highest-order capacities at the first order level. Jesus modeled this, for he shared our human nature (the human set of highest-order capacities), an exemplified them fully. As a result, he was "fully alive" as a human can and ought to be, living a life constantly in unbroken fellowship with the Father and exhibit the virtues that are part of our nature, referred to in Scripture as the fruit of the Spirit: love, joy, peace, forbearance, kindness, goodness, faithfulness, gentleness, and self-control (Gal 5:22–23). Because we share this very same human nature at the highest-order level, we too can live in increasing conformity to how Jesus lived as these properties are increasingly exemplified in our lives in various ways.

---

11. Willard, *Renovation of the Heart*, 28, 209 (italics in original).

But this process of increasingly exemplifying these capacities at the first-order level is often blocked by temptations to sin. As Willard says, "We are free *from* sin [at the highest-order level] even if not yet free *of* it [at the first-order level]."[12] In this too, Jesus' life gives us hope. Being fully human himself, he too encountered temptations to block the first-order exemplification of these properties. He was "tempted in all things as we are" (Heb 4:15a). Yet he was "without sin" (Heb 4:15b). Jesus showed that we, too, can overcome these temptations and allow these highest-order capacities to be increasingly expressed in our lives.

For Willard, the spiritual disciplines help us overcome these challenges and so increasingly live according to our nature at the first-order level. Referring to this hierarchical structure of our soul's capacities Willard writes:

> I will not be able "on the spot" [at the first-order level] to do the good thing if my inner being is filled with all the thoughts, feelings, and habits that characterize the ruined soul [blockages in mental, emotional, and volitional capacities].... If I intend to obey Jesus Christ, I must intend and decide to become the kind of person who *would* obey [higher-order choices that lead to first-order exemplifications]. That is, I must find the [higher-order] means [disciplines] of changing my inner being until it is substantially like his, pervasively characterized [at the first-order level] by his thoughts, feelings, habits, and relationship to the Father.[13]

And as we do so, we find that we have greater intimacy with God and more consistently enjoy the fruit of the Spirit in our lives, as they are exemplified at the first-order level. Willard summarizes, "Doing what is good and right becomes increasingly easy, sweet, and sensible to us as grace grows in us."[14] As a result, God is more fully honored through our lives and we experience greater flourishing day by day.

Yet these disciplines require hard work. As Willard says, "We can lay down as a rule of thumb that if it is *easy* for us to engage in a certain discipline, we probably don't need to practice it."[15] Rather, we are to "work out our salvation" (Phil 2:12), including the hard work of practicing the spiritual disciplines. Willard observes, "We are saved by grace, of course, and by

---

12. Willard, *Spirit of the Disciplines*, 115.
13. Willard, *Renovation of the Heart*, 90.
14. Willard, *Spirit of the Disciplines*, 115.
15. Willard, *Spirit of the Disciplines*, 138.

it alone, and not because we deserve it. That is the basis of God's acceptance of us. But grace does *not* mean that sufficient strength and insight will be automatically 'infused' into our being in the moment of need."[16]

This fact leads to his understanding of the process of spiritual formation and flourishing: VIM. We first must have the desire to develop intimacy with Christ (*Vision*). On the basis of this desire we can then choose to seek him (*Intention*). Finally, only once this choice is made will we have the motivation to engage in the hard work of practicing spiritual disciplines in order to achieve this goal (*Means*).

However, by shifting the focus of spirutal formation from the soul to the brain, neurotheologians sidetrack believers from doing this important yet challenging work of spiritual formation via spiritual disciplines. For instance, instead Wilder emphasizes what he calls the "fast track" to spiritual maturity, based on so-called "brain science." This fast track motif is a central theme of his *Renovated*, for he believes this approach must replace the slow process of forming one's soul through spiritual disciplines.

Wilder argues that the activities of the "slow track" of the brain (left-brain processes in which we understand truths about God and choose to live accordingly) entail very hard work yet are ineffective in spiritual formation.[17] He states, "When we think God's thoughts in the conscious, slow track, it takes great effort, effort that does not change our character very much,"[18] and "striving is almost always the godchild of the conscious slow track in the brain. Striving is evidence we are overly focused."[19] He believes this is the problem with Willard's model of spiritual formation: "The VIM model has ideas (slow conscious thoughts) as tools . . . [but] are intention and focus the best way to learn and transform character?"[20] He believes not.

The solution, according to Wilder, is to shift our focus to the "fast track": the "right-brain" processes where we stop thinking and striving, and just experience love and attachment with God. Based on his

16. Willard, *Spirit of the Disciplines*, 4.

17. In this fast-track and slow-track understanding of how we operate, Wilder relies on Iain McGilchrist's reductionist left-brain/right-brain explanation of our mental lives. See McGilchrist, *The Master and His Emissary*. In chapters 3 and 4, I have offered a corrective to McGilchrist's egregious philosophical errors, as they are the same ones made by Wilder and Thompson.

18. Wilder, *Renovated*, 117.

19. Wilder, *Renovated*, 172.

20. Wilder, *Renovated*, 109–10.

physicalism that reduces us to a right and left brain, Wilder explains that "Nothing in the conscious, slow track of the brain has the twelve characteristics of an attachment bond. Our best character formation has been through healthy bonds—even when we had no conscious idea how attachments worked."[21] Therefore, "transformation comes when our mind goes beyond correcting our beliefs to practicing attachment love,"[22] concluding that "willful focused attention [Willard's approach] is a weak force . . . the right strategy [attachment] will take us places that simple choices [Willard's approach] cannot."[23] In sum, we must leave behind the "great effort" of forming correct beliefs and making correct choices (what he believes are left brain activities) and simply experience intimacy with God (what he believes is a right brain activity).[24]

This puts in sharp contrast the physicalist's and holistic dualist's very different understandings of spiritual formation. As we have seen, due to his holistic dualism Willard has exactly the opposite advice for growth in Christ. Because we are a soul that has a body, Willard believes our spiritual lives develop not from our left or right brain, but "from the inside of the personality. It is . . . a spiritual matter, a matter of meaning and will, for we are spiritual beings."[25] Therefore, contrary to Wilder's belief that the right brain is the fast track to spiritual maturity, the VIM model emphasizes that our thoughts and intentions, capacities of our soul, are the path to Christ-likeness.

---

21. Wilder, *Renovated*, 119.

22. Wilder, *Renovated*, 87.

23. Wilder, *Renovated*, 127. In this way, his view is similar to the excesses of the Keswick view of sanctification, often stated as "let go and let God," which absolved the believer of any effort in one's spiritual growth. For a theological analysis of the Keswick view, see Neselli, *Let Go and Let God? A Survey and Analysis of Keswick Theology*. For a survey of this and other views of growth in Christ, see Gundry, *Five Views of Sanctification*.

24. This reduction of persons to right and left hemispheres of the brain also leads Wilder to reject Willard's emphasis on the proper ordering of *Vision-Intention-Means* for spiritual formation. Willard claims that "Not just any path we take will do. If this VIM pattern is not put in place properly and held there, Christ simply will not be formed in us." (Willard, *Renovation of the Heart*, 85). But Wilder interprets these three elements of spiritual formation in terms of the "fast track" and "slow track" of the brain. From this he concludes that, instead of a proper ordering as Willard states, the "VIM model has no direction or flow. Vision, impetus, and means interact with each other." (Wilder, *Renovated*, 176). Driven by his physicalism, Wilder misses Willard's point entirely.

25. Willard, *Renovation of the Heart*, 83.

Furthermore, while agreeing with Wilder that attachment love with God is paramount, Willard believes this is the *result*, not the *cause*, of our spiritual formation. It is achieved by our choosing and then acting accordingly, in order to develop intimacy with God (again, all capacities of the soul). This is no different from any other loving relationships of attachment—these relationships develop only *as a result* of a person choosing to pursue another and then living this out day-by-day. For Willard, loving relationships are not part of the brain and don't just happen, contrary to what Wilder claims.[26]

These differences between Wilder and Willard further illustrates a danger of Wilder's *Renovated*. He is not simply "adding to" Willard, as he claims to be doing. Rather, he is offering a very different approach to spiritual formation, based on his physicalism. My concern is that neurotheologians will dissuade believers from doing the hard soul work Willard promoted with the false promise of a fast-track pathway to spiritual maturity. People who follow this route will experience just the opposite result. They will not grow in Christ as they hope and will become discouraged, experience shame, lose hope, and even give up on life in Christ altogether. This can be avoided if we understand ourselves, as Willard does, as ultimately a soul that uses our body, including our brain, to be formed in Christ and thus to flourish.

Spiritual Formation and Our Faculties

Furthermore, we saw that these capacities of the soul are grouped into various faculties, which are causally related within each faculty and with capacities of other faculties. Understanding this rich tapestry of causal relations within and among our soul's faculties helps us identify how our thoughts are causally related to our desires and choices, and how these are also related to our emotions, actions, and relationships.

For example, I may be reading my Bible and *considering* what God has done for me (mental faculty). This results in my experiencing *love* for God (emotional faculty). Feeling this sense of love leads to the *desire* to develop even greater intimacy with God (mental faculty). Therefore, I *choose* to begin practicing the spiritual discipline of service to others, as a way to be more like Jesus (volitional faculty). I begin serving my church

---

26. The exception is the loving attachment between an infant and his or her caregiver. However, when the child develops the ability to express his or her will, the will is required to continue pursuing the relationship.

by *teaching* fourth-grade Sunday school (social faculty). This includes engaging my body in order to arrive at church each Sunday morning by 8:15 a.m. (sensory faculty). As a result, over time I sense a deeper *relationship with God* developing (spiritual faculty).[27] As I better understand what is happening "inside" myself in these various faculties of the soul I can focus on engaging various faculties more intentionally, leveraging these causal connections to develop certain habits (tendencies or dispositions to exemplify certain capacities at the first-order level). These habits will, in turn, increasingly allow my spiritual faculty to exemplify its capacity of intimacy with God at the first-order level.

Conversely, as a negative example of how our faculties can work together toward spiritual deformation or decline, I may *enjoy* (emotional faculty) being around a female colleague. I begin to *imagine* (mental faculty) having an affair with her. From this I *choose* (volitional faculty) to spend more and more time with her. I therefore arrange my schedule *to meet* her for drinks after work (sensory faculty), and I *engage her* in witty conversation (social faculty). As I continue down this path, my *intimacy with God* (spiritual faculty) decreases. Again, by understanding these faculties of the soul and the causal connections among them, I can head off beliefs, desires, emotions, choices, and actions that will negatively affect my spiritual capacities (in the words of Ephesians 4:18, that will ultimately result in hardening my heart toward the things of God).

Yet in neurotheology this rich tapestry of causal connections among capacities of the soul in its various faculties evaporates. All we are left with is an understanding of the drab firings of neurons. At best, this is of no help as we seek to love God and follow his ways. At worst, it leads disciples of neurotheologians to conclude that, if we are only a body, whatever the body (neurons) desire, the body (neurons) should have. Some may take this as a justification for not seeking to grow in Christ at all. They may claim that "I'm just wired this way" or, more crassly, "My brain made me do it." Or they may even say their lack of interest in spiritual things is "just the limitations of my left brain." If lived out consistently, this conception would lead us to follow our passions rather than the Spirit. Along the way, many spiritual lives will wither, many marriages will collapse, and God will be mocked.

---

27. Of course, these are only a few of the causal relationships that occur between the soul's various faculties in this example.

Spiritual Formation and Our Teleology

Third, understanding that our souls and bodies have a natural objective toward which they are oriented to move also helps us better understand the nature of spiritual formation. Jesus calls us to "be perfect, as I am perfect" (Matt 5:48). The Greek word used for "perfect" is *teleios*, from the noun *telos*, which means end or goal. Jesus is encouraging us to strive toward our natural end. As discussed in chapter 5, this is the full first-order exemplification of all our highest-order capacities in each faculty of our soul. As noted above, the fact that Jesus constantly exemplified this end (this *telos*, or perfection) gives us hope and encouragement that we can move toward this end as well as we "press on toward the goal for the prize of the upward call of God in Christ Jesus" (Phil 3:14), and that our endurance will ultimately lead to a "complete result" (Greek *teleioi*, Jas 1:4), though we will never fully reach it until our full sanctification upon death (1 John 1:8–9).

The fact that we have a natural end or *telos* implies that there is a natural pathway to maturity, which can be obtained unless it is blocked in some way. Unfortunately, we do all have blockages in our path. Understanding the nature of these blockages helps us identify and remove them through the practice of specific spiritual disciplines.

Some blockages are intellectual. We may not understand biblical truth, or we may fail to apply it consistently. If so, spiritual disciplines such as study, silence, and solitude may help us better understand and apply God's truth in our lives. One role of theologians, pastors, and spiritual mentors is to help us identify these blockages and overcome them by increasing our biblical understanding and application.

Our blockages may be emotional. Wounds from our past may be hindering our growth, which we must work through in order to continue growing. Spiritual disciplines such as confession, fellowship, and solitude may help us overcome these challenges. Often we need the help of a pastor or a Christian counselor to work through such issues in order to remove these blockages.

Our blockages may be relational. Perhaps broken relationships are hindering our spiritual growth. Spiritual disciplines such as confession may be necessary, as Jesus illustrates: "If you are offering your gift at the altar and there remember that your brother or sister has something against you, leave your gift there in front of the altar. First go and be reconciled to them; then come and offer your gift" (Matt 5:23–24). Other helpful disciplines may include corporate worship, service, fellowship, and submission. Again, pastors or counselors can often help us know how best to remove these blockages and encourage us to take the first step.

Our blockages may be volitional. We may be consciously choosing to believe or do something that hinders our spiritual growth. Scripture contains many encouragements to make wise decisions that will lead to maturity. Disciplines such as fasting, solitude, sacrifice, and prayer may be of help in removing volitional blockages. Having friends who hold us accountable to following through on decisions we have made is often very helpful in staying the course and consistently making right choices.

Yet once again, if we embrace the physicalism of the neurotheologians, this all fades away. Material objects do not have a nature, and so there is no end toward which they naturally strive. Gone is any sense of an objective understanding of maturity, as well as how to overcome specific challenges through appropriate spiritual disciplines.

I have tried to show that only by better understanding the nature, goodness and value of both our souls and bodies can we understand how both are integral to our spiritual formation. With this understanding, we are well on our way to flourishing as human beings and in our walk with Christ.

In addition to helping us better love God, a proper understanding of what we are also helps us love others as they truly are. The final chapter will consider how understanding holistic dualism helps us love others in our daily lives and in our specific professions.

10

Soul, Body, and Loving Others

Some of the claims of reductionist science are not only conceptually incorrect or even unintelligible, they have major social implications. The words we use, the concepts by which we analyze and present biological discovery, deeply affect the way in which we see ourselves as human beings.
—M. R. BENNETT AND P. M. S. HACKER[1]

If we want to know what it is to save a human being, to redeem the human soul . . . we cannot find a better way to begin than by asking: what did God make when he made us, and how could creatures such as we be at risk and at loss?
—DALLAS WILLARD[2]

> **CHAPTER SUMMARY**
>
> This final chapter considers how we can better love others based on understanding ourselves as souls deeply united to our bodies. I discuss three ways in which this understanding, in contrast to the physicalist and Platonic/Cartesian options, better helps us love others

---

1. Bennett and Hacker, *Philosophical Foundations of Neuroscience*, xiv.
2. Willard, *Spirit of the Disciplines*, 45.

> in need. First, it helps us make sense of and gives us motivation for evangelism and mission to a hurting world. Second, it helps us care for "the least of these" who may be harmed due to their biological (functional) limitations. It also provides an objective rationale to work toward justice for all. Finally, I discuss how holistic dualism helps us love others through our professions.

OUR UNDERSTANDING OF WHAT people are will determine how we can help them flourish. It will help us best love others as ourselves. We see this in practical terms as we consider the nature of evangelism and missions, biomedical ethics, working toward justice for all, and loving others through our professions.

LOVING OTHERS AS CHRIST'S AMBASSADORS

Evangelism and Missions: Proclaiming the Good News

In Search of the Soul, a book exploring various understandings of what we are, begins by listing a number of important questions that every formulation of our nature must answer: "How should we understand 'salvation'? What needs to be 'saved'? . . . How ought the church to be extending itself in mission? Mission to what? The spiritual or soulish needs of persons? Society at large? The cosmos?"[3] Only holistic dualism offers compelling answers to these important questions.

The Two Extremes

If we focus too much on the soul and discount the body, we can easily perpetuate Gnostic ideas of ministry that focus on only saving souls and not also meeting our hurting world's physical needs. As we have seen, the body is central to who we are, and it has real needs. Meeting other people's physical needs is a very powerful way we can love our neighbors.

---

3. Green, *In Search of the Soul*, 9.

As Jesus says, "Come, you who are blessed of My Father.... For I was hungry, and you gave Me something to eat; I was thirsty, and you gave Me something to drink" (Matt 25:35).

On the other end of the continuum, if we discount the soul and believe we are only or fundamentally a body, as Thompson and Wilder do, equally troublesome consequences result. First, if people—who *appear* to have an immaterial dimension—are really just physical, why believe that *anything* exists that is not physical? The existence of objective moral values becomes increasingly implausible. Our culture will find it harder to believe that angels and God Himself exist if they are not physical, or that these supernatural beings can think, feel, desire, and make choices if they do not have brains. We must not contribute to such cultural skepticism of immaterial things. Otherwise, in the words of Os Guinness, we become our own gravediggers.[4]

Furthermore, if we are just physical beings, there is no such thing as sin. Sin involves wrong attitudes, thoughts, and choices, which lead to wrong actions. But attitudes, thoughts, and choices are properties of a soul, not a body (as discussed in chapters 3, 4, and 5). Therefore, if no soul, then no sin. Yet if there is no sin to alienate us from God, there is no need for a Savior to pay our penalty for sin and restore us to a right relationship with God.

In fact, if the physicalists are right and there is no soul (that is, no individuated *human nature*, as discussed in chapter 5), there is no human nature that Jesus could take on in his incarnation, making him truly "like us in every way" (Heb 2:17). Yet if this is the case, he cannot be an *equal* substitute for us in his death.[5]

Finally, if the physicalists are correct and we have no reason to believe in sin, Jesus' incarnation, or Jesus as a suitable sacrifice for our sin, there is no good news of the salvation of our souls. Sharing such a message is, at best, a waste of our and other's time. At worst, for the physicalist, sharing this message is actually harmful, as it diverts attention from the real physical needs of people all around us to the concerns of an imaginary soul. Thompson seems to nod in this direction, downplaying the emphasis on personal salvation in the Bible. He states, "The

---

4. Guinness, *The Gravedigger File*.

5. Concerning the incarnation, see Van Horn, "Dualism Offers the Best Account of the Incarnation," 440–51. For a physicalist response, see Merricks, "The Word Made Flesh," 452–68. Concerning substitutionary atonement, see Wallace, *Aiding the Christian Scholar*, 167–68; DeWeese, "One Person, Two Natures," 144–53.

whole of Scripture points to the idea that God is not first and foremost intending to save us as individuals."[6] He goes on to explain that instead God desires to "save" the larger social context from which we arise and have our being.[7]

But this is clearly false. Scripture speaks often of the salvation of our souls, as discussed in some detail in chapter 2. Jesus warns us to fear him who can destroy both our bodies and our souls (Matt 10:28), and he promises the thief dying next to him that they will soon be together in paradise, though their bodies will be dead (Luke 23:43). This message that souls can be saved is what we are to share "to the remotest part of the earth" (Acts 1:8), because "as many as received him, to them he gave the right to become children of God, even those who believe in his name" (John 1:12). The physicalism of Wilder and Thompson, with these logical implications for evangelism and missions, is not a way to love our neighbors.

The Soul-Body Synthesis

The middle way of holistic dualism avoids these errors and implications. In affirming the value of our body, it avoids any idea that meeting physical needs is not an important part of loving our neighbors, consistent with how Jesus acted repeatedly in his own ministry. On the other hand, by providing a robust understanding of the soul, holistic dualism also avoids the severe consequences of physicalism discussed above. We can help others make sense of our shared sinfulness and need of salvation. Again, this is what Jesus modeled as he constantly proclaimed the good news concerning the salvation of our souls (such as in John 3).

Holistic dualism also makes sense of what it meant for Jesus to take on "the very nature of a servant, being made in human likeness" (Phil 2:7). By understanding what a nature is in general, and what human nature is in particular, we can more fully grasp what it means for Jesus to have the *very same* nature as us—this human-specific set of highest-order capacities. As a result, we can see how, being truly and fully human, he could be an *equal* sacrifice for our sins.[8]

---

6. Thompson, *Anatomy of the Soul*, 99.
7. Thompson, *Anatomy of the Soul*, 99–100.
8. In addition to being an equal sacrifice for our sins due to his humanity, Jesus was also an adequate sacrifice for our sins, due to his deity. He was *truly* a human, but not *only* a human.

In these ways, holistic dualism best helps us make sense of the gospel, and therefore it enables us to properly explain the logic of the gospel when non-believers ask us what sin is, in what sense Jesus became one of us, or how he could die in our place. This allows us to most fully love our neighbors across the street and around the globe.

Biomedical Ethics: Fostering True Human Flourishing

The Cartesian Dualist and Physicalist Errors

We also love our neighbors by helping them live full and healthy lives. But what makes a full and healthy life will depend on what persons are understood to be. If we are really just a soul that happens to have a body, we will define a "full and healthy life" primarily (or exclusively) in terms of the soul. One way we see this anthropology expressed is in the belief that the soul, through faith, can have full control over the body. This view often claims that if a person truly believes he is well, the body *must* respond accordingly (since the causal chain runs only from soul to body). People who believe this often deny being ill and instead claim "by faith" that God has brought healing, regardless of how they actually feel or what the doctors say. Therefore, as I've personally witnessed several times, medical (physical) treatment for the illness is refused, which often leads to further complications and sometimes even death.[9] Promoting a view of the human person that so devalues the body is not a way to love our neighbor.

At the other end of the spectrum, reinforcing the view that people are ultimately physical beings does not lead to loving our neighbors either. For instance, without a nature that defines what we are, our value can be determined only by our body's abilities or physical characteristics. Depending on what abilities or characteristics are selected to be most valuable, some persons are thus deemed more valuable, or even more human, than others. Several examples will be offered below to illustrate how easily this error leads to harming rather than loving our neighbor.

---

9. We see this in the "Word of Faith" movement, of which the late Kenneth E. Hagin was the father and leading proponent. For instance, he states, "When the Bible talks about suffering, that doesn't mean 'sickness'. We have no business suffering sickness and disease, because Jesus redeemed us from that." Hagin, *Must Christians Suffer?*, 2. In the more extreme case of Christian Science, the physical body is denied altogether: "The physical body is a manifested false sense of being." Barratt, "The Body and You."

The Holistic Dualist Solution

Again, holistic dualism avoids the errors of both these extremes. The body is taken seriously, and so it must be respected in its own right. We should trust medical professionals, even if they tell us what we don't want to hear or believe. On the other hand, the soul is also taken seriously. Our shared human nature (discussed in chapter 5) defines objectively what we are and grounds the ethical treatment of all people, regardless of any abilities or other characteristics a person may or may not exhibit.

Case Study: The Ethics of Abortion[10]

The debate over whether abortion is morally appropriate illustrates how these views of what it means to be human lead to very different conclusions. The primary question when considering the morality of abortion is when the person begins to exist. Before this point, removing a mass of cells in the woman's body is morally acceptable, for it is not a person, and therefore innocent life is not being taken. Yet after life begins, there is more in the woman's body than just a mass of cells—there is another person. At this point, the procedure results in the taking of an innocent life and is therefore morally wrong, for we should not use others as we wish and without their consent in order to achieve our goals.

As we have seen, the Cartesian view holds that our bodies are machines that run on their own. Therefore, the fetus becomes a person only when the soul is added. From this, it follows that if the soul has not yet been "implanted" in the body, abortion is morally appropriate. Given this view of what a person is, it becomes crucial to determine when the soul is "added" to the body. Yet for the Cartesian dualist, it is hard if not impossible to answer this question. If the body is capable of running on its own steam whether or not it is ensouled, no bodily function could be identified as indicating that a soul is present. As a result, Cartesian dualism gives us no information about when a medical procedure does or does not result in the taking of innocent life.

On the other hand, since physicalists believe we are ultimately a mass of cells, life can be defined only by what these cells can or cannot

---

10. Other biomedical ethics issues to which these considerations apply include genetic technologies, human cloning, and euthanasia. For an excellent treatment in relation to a wide range of ethical issues, see Rae, *Moral Choices*. Concerning euthanasia, see also Moreland and Wallace, "Aquinas vs. Locke and Descartes."

do. If certain functions are *absent*, the cells are not a human person, and so their removal is morally appropriate, especially if other goods result, such as relieving a mother's mental or emotional distress. Yet if these functions are *present*, the cells are a human person and abortion is not morally appropriate.

However, there is no agreement among physicalists as to which function or functions define the threshold of life. This raises the sticky problem of who gets to determine which functions count. Many argue for various stages of development in the womb, such as implantation in the uterus wall, detectable brain activity, or viability (the ability to live outside the womb), in which case abortion is permissible up to that point of development. Yet other physicalists argue that the key function conferring personhood is the ability to have an awareness of itself, its environment, and its future, among other things. These functions don't develop until several months after the child's birth, and so infanticide is also morally appropriate until these functions are expressed.[11] There is no clear criterion by which to evaluate these competing claims. In fact, a physicalist could defend other, even more advanced functions as the threshold of personhood, leading to the treatment of people with limitations in those functional areas as non-persons. These logical implications of physicalism certainly do not foster love of neighbor!

Holistic dualism offers a solution to these conundrums of Cartesian dualism and physicalism. It provides philosophical reasons to believe that both the soul and the body come into existence at the very moment of conception (see chapter 6). Therefore, when there is a functioning body (even at its earliest stages of development), a soul (an individuated human nature—a person) is present as well. This is true whether or not the person can express specific capacities of human nature at various points in his or her development. These capacities are, in virtue of being a human person, always present at the highest-order level, and the child will naturally grow to express these abilities at the first-order level if not hindered (see chapter 5).

Therefore, holistic dualism is superior to Cartesian dualism in providing sound philosophical reasons to justify the claim that life begins at conception. It is also superior to physicalism in avoiding the slippery slope of determining a person's value based on one's particular abilities.[12]

11. For instance, see Singer, *Practical Ethics*, 160. See also Tooley, "In Defense of Abortion and Infanticide," 209–33.

12. See Moreland and Rae, *Body and Soul*, 231–60. For discussion of related issues,

Social Ethics: Pursuing Justice for All

What we understand people to be also has wide-ranging implications for how we believe society best functions. Platonic/Cartesian dualism tends to encourage withdrawal from cultural issues so as to focus more strongly, or even exclusively, on the spiritual needs of others. This is because it rests on Plato's view that the immaterial realm is far more important than the physical realm. For Cartesian dualists, working for social change is like polishing the handrails of the Titanic—it is ultimately a waste of time. It is much more important to get people into life rafts (i.e., to meet their need to be spiritually rescued). As a result, in the cases described below, Platonic/Cartesian dualism won't have much to offer in response to questions of social engagement.[13]

As discussed above regarding biomedical ethics, the physicalist must ground what it is to be a person, and therefore a person's value and rights, in one's body—one's physical characteristics or abilities. Therefore, for the physicalist, there logically can be no such thing as shared intrinsic value, fundamental equality, justice for all, or inalienable human rights. There is literally nothing we all share in common that could provide a foundation for these values.[14]

Rather, a person's value must be derived from one's functional abilities or one's place in the social order. Thompson explicitly rejects the claim that one's value is only in virtue of his or her nature, emphasizing instead the role of one's community in conferring value: "A person's inherent value is not to be understood merely in terms of his or her individual existence . . . apart from that person's place in a larger community."[15]

---

see Moreland and Rae, *Body and Soul*, 263–84, 287–312; Rae, *Moral Choices*, 126–200.

13. This is certainly not true of all Cartesian dualists in actual practice. However, this position follows logically from the Platonic or Cartesian view, which sees the immaterial and material realms as very loosely connected and the spiritual realm as significantly more important. For more on various views of Christianity and cultural engagement, see Niebuhr, *Christ and Culture*; Crouch, *Culture Making*.

14. Again, I am not claiming that physicalists never promote such values. My point is simply that doing so is inconsistent with what the physicalist takes to be real.

15. Thompson, *Anatomy of the Soul*, 240. He cites 1 Corinthians 12 and 13 in support. Yet Paul here is discussing the value we bring to and receive from being in healthy relationships with others as each of us exercises our various gifts for the common good. Paul's focus is on our instrumental value in the community, not our intrinsic value as a person. This is very different from Thompson's emphasis on one's *inherent* value being tied to one's relationships or community.

But if our value is tied in any way to something other than our fixed, unchanging nature, this opens the door to treating others differently based on their relational or functional differences. As a result, those with less of the "right" characteristics or abilities are often discriminated against. In more extreme cases, they are treated *inhumanely*—literally as non-humans. An extreme example was the horrors of the Holocaust, where personhood, value, and rights were defined in terms of the characteristics of one's ethnicity.[16] Regardless of how extremely one applies these implications of physicalism, this certainly does not lead to loving one's neighbor.

Holistic dualism has a philosophical foundation that supports the full equality of all people due to the intrinsic value of each person's soul, which possesses the image of God.[17] This shared human nature is the objective basis for treating one another equally and in a humane way. Therefore, no matter how different we may be in other physical and functional ways (such as ethnicity, gender, socio-economic position, religious practices, sexual orientation, and so on), each person deserves to be treated with the utmost respect. This is true even though we may deeply disagree with one another on any number of important issues.

Not only does holistic dualism provide an objective basis for protecting each individual's rights, but its commitment to our shared humanity best promotes the common good as well. No one group can claim special rights due to a physical or functional distinctive such as ethnicity, gender, religion, or sexual orientation. We are all at the most fundamental level *human*, and only by focusing on what we all share in common can we foster the flourishing of *all* humans—the *common* good. This viewpoint counters attempts to privilege one group or class over others based on differences that are not grounded in our shared human nature. It prevents our society from becoming fractured into divisions that ultimately undermine the *common* good by promoting the *particular* goods of those groups that have power at any given time.[18] Holistic dualism prevents

---

16. Before this implication of physicalism played out in World War II, it was identified by Edmund Husserl (1859–1938) in his famous lecture before the Vienna Cultural Society on May 7 and 10, 1935, with the original title "Philosophy and the Crisis of European Man" (available online at http://www.markfoster.net/struc/philosophy_and_the_crisis_of_european_man.pdf, accessed December 28, 2023). His prophetic words fell on deaf ears. He later expanded these ideas in his *The Crisis of the European Sciences and Transcendental Phenomenology*.

17. The Cartesian dualist affirms this as well, once the soul is implanted in the body.

18. For a very insightful evaluation of this loss of a sense of our shared humanity

this and instead fosters the common good by offering good reasons to ground rights in a shared human nature that determines how we flourish. In this way we can best love our neighbor as ourselves.

LOVING OTHERS THROUGH OUR PROFESSIONS

The last aspect of loving our neighbors that I will discuss involves promoting the good of others through the work we do in our professions. How we understand the relationship of our work to the good of others depends on what we believe others to be and how they (and we) flourish. Obviously, I cannot cover every occupation here, so I have selected just a few examples. Even in these cases, much more can be said. My goal is to illustrate briefly how holistic dualism leads to our best loving our neighbor as ourselves through our work. (As I have noted previously, people often live inconsistently with what they say they believe. Therefore, there are certainly physicalists who engage their professions as if they are dualists, and dualists as if they are physicalists. My point is that, in as far as people live according to their professed anthropology, dualists best love others through their work.)

Overall Approaches to Work

For those who lean toward the Platonic/Cartesian extreme that the material realm is of secondary importance or even inherently evil, loving others is often viewed exclusively in terms of activities in church or other "ministry" contexts that care directly for the soul. One's professional life is, at best, only instrumentally valuable as a way to share the gospel with colleagues or earn money to support one's family, church, and missions. But work itself is not seen as a means to advance God's Kingdom on earth. As a result, little thought is given to the value of work as intrinsically related to loving one's neighbor and human flourishing.

On the other hand, if people are material beings, it follows that we best serve others in our professions by meeting their physical needs. Making others comfortable and (physically) healthy should be the focus. Insofar

---

and the common good, of *expressive individualism* as its replacement, and of the deeply problematic results, see Trueman, *The Rise and Triumph of the Modern Self*. For an equally good but less academic treatment, see Trueman, *Strange New World*. See also Lee and George, *Body-Self Dualism in Contemporary Ethics and Politics*.

as physicalists can fulfill those desires, their work contributes to human flourishing. Yet as the examples below will illustrate, their approach does not ultimately guide us toward loving our neighbor most fully.

Between these extremes, again, stands holistic dualism. As affirmed by a view of reality that appears in Scripture, and is echoed by philosophers from Aristotle on, the material and immaterial realms are both extremely valuable. This applies not only to souls and bodies, but also to the immaterial and material dimensions of the world around us. It implies that our work is more than just what we do to get a paycheck. It is a place where we can help others flourish by seeking to meet the needs of both body and soul. This most fully allows us to love our neighbors as ourselves. I'll illustrate with reference to eight professions.[19]

Loving Others Through Education[20]

To what end should we educate? If people are fundamentally physical, the goal is to teach students how to make a good living, so that they can meet their own and other people's physical needs. This perspective will lead to a heavy emphasis on science, technology, engineering, and mathematics (STEM), for these fields meet other people's physical needs by developing or further implementing technology necessary for "creature comforts" (and they also provide the best-paying jobs). But as Maslow observed in his hierarchy of needs,[21] we are not fulfilled when only our physical needs are met. We crave more.

If, on the other hand, we also have a soul, the goal of education is to provide the "more" that we crave. STEM courses are important, for we do have physical needs. But so are studies in the humanities (such as history, philosophy, and literature), which focus more directly on the immaterial realm by discussing the nature of truth, morality, the human person and condition, God, and so much more.

---

19. If I did not include your vocation as an example, I invite you to ask these same questions about your own field: What are the implications of my work if people are purely or fundamentally physical and therefore have only material needs? What are the implications if those I serve are also immaterial and therefore also have needs in the various faculties of their souls, such as emotional, spiritual, and social needs?

20. See Spears and Loomis, *Education for Human Flourishing*; Habl, *On Being Human(e)*.

21. Maslow, "A Theory of Human Motivation."

Loving Others Through Medicine[22]

If we are only (or fundamentally) a body, what ails a person can only (or most fundamentally) be physical. Therefore, reading the patient's chart tells you all you need to know. There is no need to talk with patients to get their "first-person" perspective (which, as discussed in chapter 3, does not exist for the physicalist). Furthermore, treatment should be focused on physical interventions, such as medication, surgery, or physical therapy, for that is the only (or at least the ultimate) source of the problem. For example, recently I went to see a specialist for a medical issue. The physician hardly looked at me. He read my charts, looked through my medical history, and told me what he was going to do. This did not communicate love and respect to me, but just the opposite.

On the other hand, the holistic dualist recognizes that although the medical issue might be purely physical, it might also involve the person's immaterial dimension as well (recall the findings on how best to treat Social Anxiety Disorder, mentioned in chapter 6). Understanding how the patient is feeling "inside" by asking for a first-person perspective becomes as important as the medical tests that can be run "outside" such as blood work and MRI scans. This communicates great love and respect to the patient. Furthermore, the physician who affirms holistic dualism is open to the possibility that the presenting physical symptoms may *ultimately* be caused by an ailment of the soul, such as anxiety causing an ulcer. On the other hand, a holistic dualist counselor may find that a presenting mental disorder is ultimately caused by a physical condition (such as poor sleep habits leading to extreme mood swings). One good example is the work Jeffrey Schwartz and Sharon Begley have done in treating obsessive-compulsive disorder (OCD) and other mental disorders based on ideas drawn from holistic dualism.[23] In these ways, medical professionals love others best by caring for the person as an incarnated soul.

---

22. See Pellegrino and Thomasma, *Helping and Healing*; May, *The Physician's Covenant*.

23. Schwartz and Begley, *The Mind and The Brain*. For an illuminating, very personal account of the interaction of body and soul in overcoming anxiety, see Moreland, *Finding Quiet*. See also Rickabaugh and Moreland, *The Substance of Consciousness*, 337–39.

Loving Others Through Business[24]

If we are physical beings, then we can best love others by using our business expertise to meet their physical needs (as well as our own). Success is then measured by material compensation—the financial bottom line. There would be no higher goal in business.

On the other hand, if you believe that people are both body *and* soul, truly loving others will mean helping them flourish in both dimensions. For instance, you may choose to prioritize an employee's mental and relational health (the health of his soul) by not having him work late and instead sending him home to be with his family, rather than trying to improve the company's profits by asking him to work more hours. This is sometimes referred to as seeking the "triple bottom line"—i.e., focusing on not only financial health but also social and environmental health in order to foster the flourishing of employees and everyone else the company influences.[25]

Loving Others Through Architecture[26]

For the physicalist, all needs are physical, and therefore those who design spaces where people live and work need be concerned only with the space's physical utility. Yet the dualist must also consider how to include beauty, symbolism, and even whimsy in the design, in order to nourish the soul.

I recently visited Timisoara, Romania. Along one side of the main promenade are houses and shops built prior to the Soviet era. The architecture is beautiful, including artistic elements that don't serve any physical need, but only express the designer's creativity for the enjoyment of others, and perhaps communicate theological truths or other important ideas. On the other side of the boulevard are houses and shops built after Soviet occupation. The architecture is bleak, void of any artistry, and purely utilitarian. Clearly, it was designed to meet physical needs and nothing more. If we are only physical beings, this would be enough. But

---

24. See Wong and Rae, *Business for the Common Good*.
25. See Michael, "Dualism at Work," S41–S69.
26. See Morris, *If Aristotle Ran General Motors*, which is also helpful in a range of other business-related issues. I use here the work of architects, but the same is true for interior designers, city planners, and all others who design spaces for a living.

everyone wanted to live and work on the artistic side of the street, for it enlivened their souls. This is true because holistic dualism is true.

Loving Others Through Law and Politics[27]

Chapter 3 discussed why there can be no free will if physicalism is true. In this case, political science becomes just as physicalistic as biology, dominated by fixed laws that determine how citizens can be directed toward the state's ends. Marxists have worked harder than anyone to apply this physicalistic philosophy in their statecraft. And it has not led to human flourishing.

In contrast, if we also have a soul, we have free will and so there is no pure "science" of politics. As political commentator Peggy Noonan recounts,

> The other day . . . a student . . . asked . . . about the predictability of human response to a given set of political stimuli. I answered that if you view people as souls . . . then nothing political is fully predictable, because you never know what a soul will do, how a soul will respond, what truth it will apprehend and react to.[28]

Therefore, for the holistic dualist, political "science" should develop and implement a rule of law that fosters opportunities for citizens to flourish by expressing *all* they are, in body and soul, such as ensuring that they have freedom to learn and to engage others in healthy relationships.[29] This is how we best love our neighbors.

Loving Others Through Science[30]

When it comes to the hard sciences, the physicalist starts with the epistemological commitment that only what can be known through science can be true (chapter 7). Therefore, science must limit itself to explanations that involve only material and efficient causes (chapter 8).

---

27. See Beckwith, *Politics for Christians*.
28. Noonan, "We're More Than Political Animals," A5.
29. Especially helpful in thinking about law from a Christian perspective is Schutt, *Redeeming Law*.
30. See Plantinga, *Where the Conflict Really Lies*.

Yet this is ultimately harmful to others, for eventually they will see through this reductionism and lose faith in the good of science. This is the worry of Bennett and Hacker:

> Neuroscientists are understandably eager to . . . share with the educated public some of the excitement they feel about their subject. . . . But by speaking about the brain thinking and reasoning, about one hemisphere knowing something and not informing the other, about the brain making decisions . . . and so forth, neuroscientists are fostering a form of mystification and cultivating a neuromythology that are altogether deplorable. . . . Once the public become disillusioned, they will ignore the important genuine questions which neuroscience *can* both ask and answer. And this surely matters.[31]

The holistic dualist is not beholden to this philosophy of the secular age. We don't need to try to explain everything scientifically. We best love others by being honest about what science does and does not know, as well as what science can and cannot know.

Furthermore, if science is ultimate and unquestionable by other disciplines, ethicists cannot raise concerns about possibly problematic ethical implications of scientific inquiry or application. As I was once told when participating in a panel discussion on the morality of reproductive technologies, "If science *can* do it, science *should* do it!" But scientists who affirm that reality also exists outside the physical realm, and therefore that truth can be found in other fields, have good reason to limit scientific undertakings that could have negative and far-reaching consequences (such as some applications of genetic engineering). In these ways, scientists who are holistic dualists can best love their neighbors.

Loving Others Through Computer Science[32]

For the physicalist, our thinking is nothing more than electrical impulses moving along neural networks. Furthermore, many physicalists define our ability to think at a sophisticated level as that which makes us human and therefore deserving of human rights. Yet if this is what thinking is, computers also think, for they too consist of electrical impulses moving

---

31. Bennett and Hacker, *Philosophical Foundations of Neuroscience*, 409.
32. See Thacker, *The Age of AI*. Understanding what we are has much to say about related issues. See Rickabaugh and Moreland, *The Substance of Consciousness*, 339–43.

along networks of wires. Therefore, human thinking is different only in *degree*, and not in *kind*, from that of a computer. If this physicalist account is true, then as computer technology continues to develop and a computer's computational sophistication eventually equals or even exceeds that of human beings, at that point the computer will have met the functional definition of a human person and will be entitled to human rights. Moreover, as their functional abilities advance beyond ours, they could qualify for greater human rights than we have!

In contrast, the holistic dualist maintains that our thinking is different *in kind* from that of a computer (as an activity of the soul, not of the brain, as discussed in chapter 3). Therefore, computers will never be able to *think* or exhibit any other aspects of the soul—of being a person. Accordingly, we have rights, as persons, that even very sophisticated computers should never have.[33] In this way, we can justify loving our neighbors more than our machines, regardless of how our neighbor's functional or intellectual abilities compare to those of our computers, robots, or AI simulations.

Loving Others Through Vocational Ministry

Finally, I offer a few words for those called to the pastorate, the mission field, or vocational ministry in other contexts.[34] I'll speak in the first person, as this is my calling as well.

---

33. Similar reflections refer to animal rights. If we and animals are both machines, our value, worth, and thus rights should be equally based on functional abilities. Therefore, if an animal functions better in a way determined to be fundamental to personhood than a human person does, the animal's rights should trump the human's rights based on physicalism, as philosopher Peter Singer argues in his popular *Practical Ethics*. However, if we are a body and a soul, and if our soul, bearing the image of God, ultimately gives us intrinsic worth and value, then human rights are paramount, even if an animal functions better than a human in some way. Of course, according to the Christian worldview, animals are part of God's good creation that he charges us to care for in Genesis 1:26–28. So animals certainly should be treated as living creatures precious to God and therefore with some rights (such as the right to adequate food and water). Yet they do not have *equal* rights as human persons, for they do not share the image of God.

34. What follows also applies to those in more informal mentoring relationships, as well as to the calling of parents to nurture our children (Prov 22:6). If our mentees or children are purely physical, our nurturing of them will pertain to their physical health and well-being. Yet if they are ultimately a soul united with a body, our role is much larger, also focusing on spiritual and character formation. Moreover, we are therefore required to teach them to love others as immaterial beings, for, as C. S. Lewis wrote, "You have never talked to a mere mortal" (*The Weight of Glory*, 45).

If physicalism is true, then there really is no "spiritual" work for us to do, because people are not spiritual beings. At best, our efforts are amusing yet irrelevant; at worst, they are harming others by promoting ideas and lifestyles that deny a person all the physical pleasures of this life, and by giving those under our care a false hope for an imaginary life after the body dies. In either case, we have nothing of value to offer.

Therefore, we must redefine "ministry" in terms of loving people's brains. Ministry training must be refocused around understanding neuroscience so as to help those under our care develop an increasingly integrated prefrontal cortex so that they can increasingly reflect the mind of Christ.[35]

Furthermore, if physicalism is true and as technology develops, one day a brain implant might be developed that effortlessly provides deep spiritual experiences of intimacy with God and full maturity in Christ, making our work completely unnecessary.[36] At that point, we pastors and counselors will be out of a job. As is occurring in so many other professions, we will be replaced by technology.

Not so if holistic dualism is true. In this case, those of us who are called to vocational ministry are attending to something that is equally real—the soul (including its relation to the body). We can boldly affirm the value of helping others in their *spiritual* formation, not *neural* formation, as an important way to love others. And even if an implant is developed that provides "spiritual" experiences, we know this would only be a pseudo-spirituality of the brain, not the real spirituality of an immaterial soul. In this way, we can show great love for our neighbors and help others do the same. To emphasize this important point, I conclude with Dallas Willard's encouragement to those of us in vocational ministry:

> The soul has very much been at the center of traditional Christianity. But in the contemporary context you will hear very little about the soul in Christian groups of whatever kind. . . . There is very little said from the pulpit about the soul as an essential part of our lives. . . . Ignoring the soul is one reason why Christian churches have become fertile sources of recruits for cults and other religions and political groups. It is not reasonable to think the soul would be properly cared for when it isn't even seriously acknowledged. . . . [We] have suffered from harmful influence

---

35. Recall again Thompson's belief that "Jesus' [brain] . . . reflects the most integrated prefrontal cortex of any human of any time." Thompson, *Anatomy of the Soul*, 180.

36. I thank Osam Temple for this insight.

by the secular intellect, which frankly abhors the soul.... We have very much lost "soul" language and are embarrassed by it—though it still breaks through in the Bible and older Christian writings and in odd places here and there in contemporary life and art.... This is not a desirable situation, to say the least, and certainly it is not compatible with the serious undertaking of spiritual formation. Our preachers and teachers must emphatically and repeatedly acknowledge the soul as the living center of Christian life that it is, and they must reassume their responsibility for the care of souls.[37]

---

37. Willard, *Renovation of the Heart*, 208.

Conclusion

Our culture increasingly assumes only what we can see is real, including the body but not the soul. We must stop aiding our culture in sliding toward physicalism. This includes standing against neurotheology, which is leading many believers to join the ranks of practicing physicalists who live as if we are just bodies or brains.

Again, I do not deny the positive value of what neuroscientists are learning. Yet we must correct the excesses of neurotheologians. To do so we simply need to remember the "both/and" principle. What we discover from neuroscience is true and good, but it is only half of the story. The other half of the story is the role the soul plays. We must chart this middle course by speaking of both the soul's role and the brain's role in thinking, desiring, believing, choosing, being spiritually formed, and so on. (See chapters 5 and 6 for more on the two-way, "both/and" causal relationship between soul and body.)

In very practical terms, this means that whenever we find ourselves in a conversation about neuroscience, we should affirm the value of these scientific discoveries. But also, perhaps even in the same breath, we must assert what we know about the soul and its role in how we engage the world. Here's a possible friendly yet provocative response: "Those are exciting discoveries! They help me understand better how we flourish as embodied souls, using our brains to engage the world around us." This will help us continue to properly articulate what we are and how we flourish by keeping the soul in the conversation. Such a comment may also provoke additional questions and thereby open the door for us to talk about spiritual things, including the gospel, as it relates to what we are. For this reason, as well as the value of our own understanding, we must continually develop a robust understanding of our nature. The suggestions for further reading in the Appendix are a good place to start.

Beyond anthropology, my hope is that this discussion will serve as a model of how to integrate biblical truth and philosophical insights with the many other issues we encounter in modern life. Doing so can help us untangle confusions and "think Christianly" about these other topics, as I have sought to do here concerning anthropology. As a result, we will be helping ourselves and others base our lives on what is true and will therefore "flourish, like trees deeply rooted and nourished by a pure, unpolluted stream, vibrant and heavy with fruit in season" (Ps 1:3, author's paraphrase).

Glossary of Technical Terms

A

Anthropology—the study of what we (human persons) are

> *Dichotomist anthropology*—understanding human beings as having ultimately two dimensions—body and soul
>
> *Trichotomist anthropology*—understanding human beings as having ultimately three dimensions—body, soul, and spirit

Argument—a series of statements (premises) that lead logically to a conclusion

Axons—the "tails" of neurons that carry electrical impulses of their associated neurons

B

Begging the question—a logical fallacy in which the conclusion is already assumed in one or more of the premises

Binding problem—the need for something to unite together, or "bind" our different experiences into our unity of consciousness. A problem for the physicalist who denies a soul exists.

C

Capacity—the ability to manifest a property. Also known as a "disposition."

> *First-order capacities*—capacities that I currently exemplify, such as the ability to do algebra
>
> *Second-order capacities, third-order capacities, etc.*—higher-order capacities needed to be able to express lower-order capacities, such as the ability to memorize formulas and the ability to think abstractly
>
> *Highest-order capacities*—the ultimate set of capacities true of all and only members of a certain type of being. Humans have a distinct set of highest-order capacities (see Human nature).

Cartesian—having to do with the thought of Rene Descartes (1596–1650)

Cartesian Dualism—see under Dualism

Cause—that which gives rise to an effect; the reason something exits or behaves as it does. Aristotle identified four causes required to provide an adequate explanation of anything in the world:

> *Material cause*—the matter of which it is composed (e.g., the wood and bricks of a house)
>
> *Efficient cause*—the energy used to bring it into existence (e.g., the carpenters who built the house)
>
> *Formal cause*—the form or essence which determines how the energy will form the matter (e.g., the blueprint for the house)
>
> *Final cause*—the reason for which it is caused to exist; its end or telos (e.g., the desire of a family to have a home)

Cerebellum—a large structure in the lower back region of the brain

Cerebral cortex—the outer layer of the cerebrum

Cerebrum—the top, outside, largest part of the brain, consisting of the brain's two hemispheres, what is usually shown in pictures of the brain

Cingulate cortex—the inside surface area between the two sides of the cerebrum, not visible in typical pictures of the brain

Communicable attributes—the attributes of God that we also possess, to varying degrees (such as love, justice, and mercy)

GLOSSARY OF TECHNICAL TERMS 183

Complementary explanations—two or more explanations of something observed in the world that differ from one another but are equally true

Constant conjunction—instances of two things or events that always occur together

Corpus callosum—a large bundle of nerve fibers that connect the two hemispheres of the brain

D

Dandy-Walker Syndrome—a deformation in which important parts of the brain do not develop

Dichotomist anthropology—see under Anthropology

Dorsolateral PFC—the upper, outer regions of the prefrontal cortex

Dualism—two distinct things existing; in anthropology, human persons having a body and soul (anthropological dualism)

Cartesian dualism—a form of substance dualism in which the soul and body are superficially united; based on the ideas of Plato and Descartes

Emergent dualism—a form of substance dualism in which the soul—a separate substance—emerges from the complexity of the brain

Holistic dualism—a form of substance dualism in which the body is caused by the soul, and therefore the two are deeply united. Variations include Thomistic hylomorphism, Thomistic-like dualism, integrative dualism, Lublin Thomism, and conditional unity views.

Property dualism—only one substance exists (the body), but it produces and sustains both physical and immaterial properties (see epiphenomena)

Substance dualism—the broader category for any view that believes we have a body and a soul (including Cartesian, emergent, and holistic dualism)

E

Efficient cause—see under Cause

Emergent dualism—see under Dualism

Enlightenment—the period of Western intellectual history broadly taking root in the 18th century

Epiphenomena—that which "rides above": in anthropology, mental events emerging from the brain when it develops a certain level of complexity

Epistemology—the sub-field of philosophy that studies how we know what is true

Essential properties (or capacities)—the highest-order properties or capacities that make us what we are and which we cannot lose

F

Faculties—a grouping of related capacities of the soul

False dichotomy—a logical fallacy in which one assumes there are only two possible answers to a problem (when in fact there are three or more), and since one option is falsified, the other option is assumed to be correct

Final resurrection—the biblical doctrine that, after our intermediate state without the body, our souls and bodies are reunited forevermore

First-order capacities—see under Capacities

Final cause—see under Cause

Formal cause—see under Cause

Functional unity—see under Unity

G

General revelation—see under Revelation

Gnosticism—the philosophy that matter is evil and spirit is good (and, beyond the scope of this book, that there is secret knowledge only some Christians have). Deriving from Plato's thought and influencing distortions of the Christian faith from early on to the present age.

Greek philosophy—philosophical ideas primarily associated Plato (c. 428–348 BC) and Aristotle (384–322 BC), though there are other, lesser-known Greek philosophers as well

H

Hemispherectomy—the surgical removal of one side of the brain

Hierarchies—the structure of capacities in which some are necessary for others to be exemplified

Highest-order capacities—see under Capacities

Hippocampus—a complex structure deeply embedded in the brain

Holistic dualism—see under Dualism

Human nature—the specific set of highest-order capacities shared by all and only human persons that make a person essentially a human

Hylomorphism—Aristotle's idea that all individual things in the world are composed of *both* form (an immaterial essence) and matter

I

Identity—cases in which two or more things share all their properties in common

Identity thesis—the view that the mind and brain are identical

> *Type-Type Identity Thesis*—a certain *type* of mental event (e.g., a memory-type event) corresponds to a certain *type* of brain event (e.g., a brain-type event)
>
> *Token-Token Identity Thesis*—a *specific* mental event (e.g., *this* specific memory of lunch two hours ago) is identical to *this* specific brain event (e.g., this specific set of neurons firing right now)

Incarnation—Jesus taking on flesh and "dwelling among us" (John 1:14)

Incorrigibility—a person cannot be wrong about what he or she is experiencing from a first-person perspective

Indexicals—statements from the first-person point of view that are used to pick us and others out as individuals, such as "I" and "you"

Individuated—each human person, while sharing with others a common human nature, is an individual being

Infanticide—the taking of a child's life after birth and before a certain functional ability is manifested

Intermediate state—the biblical doctrine that at some point our bodies will die, yet we will live on in a temporary, disembodied state until the final resurrection

L

Libertarian free will—the idea that I am free only if I could have chosen to do otherwise. If I am compelled to do something, it is not actually a choice.

Limbic area of the brain—a set of brain structures surrounding the boundary between the cerebral hemispheres and the brainstem

Logical fallacy—an instance of faulty logic

M

Material cause—see under Cause

Mereological fallacy—the logical fallacy of ascribing to the parts of a thing attributes it has as a whole, such as saying brains think rather than people think

Metaphysics—the study of what is real

Metaphysically complex—something that is complex in relation to its properties

Metaphysically simple—something that is not composed of separable parts. Also known as *mereologically* simple (from the Greek word *meros*, meaning *part*).

N

Nature—a thing's set of highest-order capacities, which it has essentially, making it what it is. Also known as the thing's natural kind or essential nature.

Negative role of philosophy—see under Philosophy

Neurons—elongated cells that carry electrical impulses. They connect with other neurons via synapses.

Neurofeedback—a type of biofeedback that measures brain waves that can help to teach self-regulation of brain function and thereby treat conditions such as anxiety

Neuroplasticity—our brain's capacity to be reshaped in order to function better

Neuroscientific hermeneutic—a method of interpreting the Bible that assumes we are essentially physical beings

Neurotheology—broadly, doing theology from the starting assumption that we are essentially physical beings. More specifically, working to develop a science of spiritual maturity based on this starting assumption.

Nonreductive physicalism—see under Physicalism

O

Ockham's razor—the principle that we should always prefer simpler explanations to more complex ones

Ontological unity—see under Unity

Ontology—the study of what is (from the Greek *ontos*, "being" or "that which is")

P

Pantheism—the worldview that understands God to be an impersonal force that is in all things

Phenomena—first-person experiences that are not reducible to third-person descriptions; for instance, the "what-it-is-like-to-hear" quality of sounds

Philosophy—the love of wisdom with the aim of living well and flourishing; from the Greek words *phileō* (love) and *sophia* (wisdom, or skillful living)

Negative role of philosophy—identifying and clarifying inadequacies of specific ideas or systems of thought

Positive role of philosophy—providing greater clarity and understanding of specific ideas or systems of thought

Philosophy of science—the evaluation of the assumptions and objectives of science (see second-order discipline)

Physicalism—the philosophical position that ultimate reality is physical

Reductive physicalism—all reality is physical. In anthropology this means mental states are identical with and completely explained by brain states.

Nonreductive physicalism—all reality is ultimately physical, but some physical things also create non-physical byproducts (epiphenomena). In anthropology, this means the brain, due to its complexity, produces not only physical but also mental properties such as beliefs. Also known as property dualism, as the brain has both physical and non-physical properties.

Positive role of philosophy—see under Philosophy

Prefrontal cortex (PFC)—the brain's highly developed frontal lobe. It is part of the cerebral cortex.

Private access—a person's direct, private, first-person access to one's own mental events such as thoughts, beliefs, emotions, and choices

Problem of interaction(ism)—the challenge of explaining how an immaterial mind can interact with a material brain

Properties—specific attributes that a thing has, such as being rational

Property dualism—see under Dualism

Propositions—the content of conscious states, which can be expressed in sentences

Proto-Gnosticism—the first-century precursor to fully developed Gnosticism

R

Relata—two (or more things) related to one (each) other

Reductive physicalism—see under Physicalism

Revelation—God's disclosure of truth to us

> *General revelation*—truths God has revealed in the created order; the domain of other fields of study, including science and philosophy
>
> *Special revelation*—truths God has revealed in Scripture; the domain of theology

S

Sanctification—the lifelong process of a Christian becoming more and more like Christ

Scientism—the philosophical belief that science is the only, or at least the best, way to know what is true

> *Strong scientism*—the epistemology that knowledge only comes from science
>
> *Weak scientism*—the epistemology that knowledge may come from other fields, but only if those findings agree with those of science

Second-order capacities, third-order capacities . . . – see under Capacities

Second-order discipline—a field of study that stands outside other fields and evaluates the disciplines' starting assumptions and objectives. Philosophy and theology are second-order disciplines.

Self-defeating—a logical fallacy in which one must assume something is true in order to show it is false, or assume it is false to prove it is true

Sensation—the felt quality of an experience, a capacity the sensory faculty of the soul. See Phenomena.

Soul –an individuated human nature; an immaterial substance

Special revelation—see under Revelation

Spiritual discipline—anything we discipline ourselves to do repeatedly in order to grow spiritually, such as fasting

Spiritual formation—the process of forming our souls to be more like Christ

Straw man fallacy—a logical fallacy in which a weak form of someone else's position is stated and then easily shown to be false

Strong scientism—see under Scientism

Substance—that which "stands under" and so unifies a thing's properties, is a unity at a time and endures through time, has a natural teleology, and is an individual thing such as a specific dog or person. Also known as a primary substance (distinguished from a secondary substance, which refers to the *type* of thing a primary substance is—its essence, such as humanness).

Substance dualism—see under Dualism

Synapses—points of contact in the brain through which electrical impulses travel

Synecdoche—a literary device in which a part is named in order to refer to the whole thing that contains that part, such as saying "all hands on deck" to mean "everyone on deck"

T

Teleology—the end toward which things naturally develop (from the Greek word *telos*, meaning end)

Thalamus—an egg-shaped mass of nerve fibers located in the middle section of the brain, just above the brain stem

Thomistic—having to do with the thought of Thomas Aquinas (1225–1274)

Trichotomist anthropology—see under Anthropology

Triune brain model—a model of the brain that divides it into the reptilian, limbic, and cortical portions

Thoughts—the content of conscious states (propositions) that can be expressed in sentences

Token-Token Identity Thesis—see under Identity

Traducianism—a view of the origin of the soul that holds it comes to be at the time of conception, transmitted in the act of reproduction

Type-Type Identity Thesis—see under Identity

U

Unity—two or more things that are, in one of two senses, one

> *Functional unity*—two things that, if divided, retain their distinctiveness yet act as one
>
> *Ontological unity*—two "things" that are actually one thing; if divided, they are divided only in how we speak of them, not in reality

V

Viability—the ability of an infant to live unaided outside the womb

VIM model of spiritual formation—Dallas Willard's core understanding of how we grow in Christ. The process begins with *Vision* (a desire to walk closely with God), followed by an *Intention* to become an apprentice of Jesus, which leads to embracing the *Means* of practicing spiritual disciplines.

W

Weak scientism—see under Scientism

Worldviews—overarching philosophies of life

Suggestions for Further Reading

ANTHROPOLOGY

Bennett, M. R. and P. M. S. Hacker. *Philosophical Foundations of Neuroscience*. Hoboken, NJ: Blackwell, 2003.

Cooper, John W. *Body, Soul, and Life Everlasting: Biblical Anthropology and the Monism-Dualism Debate*. Grand Rapids, MI: Eerdmans, 2000.

Goetz, Stewart and Charles Taliaferro. *A Brief History of the Soul*. Hoboken, NJ: Wiley-Blackwell, 2011.

Loose, Jonathan J., et al, eds. *The Blackwell Companion to Substance Dualism*. Hoboken, NJ: Wiley-Blackwell, 2018.

Moreland, J. P. *The Recalcitrant Imago Dei: Human Persons and the Failure of Naturalism*. London: SCM, 2009.

Moreland, J. P. and Scott B. Rae. *Body & Soul: Human Nature the Crisis in Ethics*. Downers Grove: IVP Academic, 2020.

Rickabaugh, Brandon and J. P. Moreland. *The Substance of Consciousness: A Comprehensive Defense of Contemporary Substance Dualism*. Hoboken, NJ: Wiley-Blackwell, 2023.

Schwartz, Jeffrey and Sharon Begley. *The Mind and the Brain: Neuroplasticity and the Power of Mental Force*. New York: HarperCollins, 2002.

Taliaferro, Charles. *Consciousness and the Mind of God*. Cambridge: Cambridge University Press, 1994.

LOVING GOD: SPIRITUAL FORMATION

Foster, Richard. *Celebration of Discipline: The Path to Spiritual Growth*. London: Hodder & Stoughton, 2008.

Johnson, Jan, et al. *Dallas Willard's Study Guide to* The Divine Conspiracy. New York: HarperCollins, 2001.

Willard, Dallas. *The Divine Conspiracy: Rediscovering Our Hidden Life in God*. San Francisco: HarperOne, 1998.

———. *Renovation of the Heart: Putting On the Character of Christ*. Colorado Springs. NavPress, 2002.

———. *The Spirit of the Disciplines: Understanding How God Changes Lives*. San Francisco: HarperOne, 1999.

LOVING OTHERS: WORK AND CULTURE

Crouch, Andy. *Culture Making: Recovering Our Creative Calling*. Downers Grove, IL: InterVarsity, 2013.

Morris, Tom. *If Aristotle Ran General Motors: The New Soul of Business*. New York: Holt Paperbacks, 1997.

Rae, Scott B. *Moral Choices: An Introduction to Ethics*, 4th ed. Grand Rapids, MI: Zondervan, 2018.

Ryken, Leland. *Redeeming the Time: A Christian Approach to Work and Leisure*. Grand Rapids, MI: Baker, 1995.

PHILOSOPHY (GENERAL)

Copleston, Frederick. *A History of Philosophy*. New York: Image, 1985.

DeWeese, Garrett J. and J. P. Moreland. *Philosophy Made Slightly Less Difficult: A Beginner's Guide to Life's Big Questions*. Downers Grove, IL: IVP Academic, 2005.

Moreland, J. P. and William Lane Craig. *Philosophical Foundations for a Christian Worldview*, 2nd ed. Downers Grove, IL: IVP Academic, 2017.

Sire, James. *The Universe Next Door: A Basic Worldview Catalog*, 6th ed. Downers Grove, IL: IVP Academic, 2020.

PHILOSOPHY OF SCIENCE

Connell, Richard J. *Substance and Modern Science*. Notre Dame, IN: University of Notre Dame Press, 1988.

Moreland, J. P. *Scientism and Secularism: Learning to Respond to a Dangerous Ideology*. Wheaton, IL: Crossway, 2018.

Plantinga, Alvin. *Where the Conflict Really Lies: Science, Religion, and Naturalism*. Oxford: Oxford University Press, 2011.

Bibliography

Amen, Daniel G. and Lisa C. Routh. *Healing Anxiety and Depression: Based on Cutting-Edge Brain-Imaging Science*. New York: Penguin, 2003.

Aquinas, Saint Thomas. *On Spiritual Creatures*. Translated by Mary C. Fitzpatrick. Reprint, Medieval Philosophical Texts in Translation. Milwaukee, WI: Marquette University Press, 1949.

———. *Commentary on Aristotle's De Anima*. New Haven, CT: Yale University Press, 1951.

———. *Questions on the Soul*. Translated by James H. Robb. Reprint, Medieval Philosophical Texts in Translation. Milwaukee, WI: Marquette University Press, 1984.

———. *Summa Contra Gentiles: Book Two: Creation*. Translated, with an introduction and notes, by James F. Anderson. Notre Dame, IN: University of Notre Dame Press, 1992.

———. *Summa Theologica*. Translated by Fathers of the Dominican Province. 1947. Reprint, Benziger Bros. Notre Dame, IN: Christian Classics, 1981.

Aristotle. *On the Soul*. In *The Complete Works of Aristotle: The Revised Oxford Translation*, vol. 1. Bollingen Series 71:2, edited by Jonathan Barnes. Princeton, NJ: Princeton University Press, 1984.

———. *Meteorology* IV.12: 389b31–32. (vol. 1). In *The Complete Works of Aristotle: The Revised Oxford Translation*, vol. 1. Bollingen Series 71:2, edited by Jonathan Barnes. Princeton, NJ: Princeton University Press, 1984.

———. *Metaphysics* Z11, 1036b25–33 (vol. 2). In *The Complete Works of Aristotle: The Revised Oxford Translation*, vol. 2. Bollingen Series 71:2, edited by Jonathan Barnes. Princeton, NJ: Princeton University Press, 1984.

Audi, Robert, ed. *The Cambridge Dictionary of Philosophy*, 3rd ed. Cambridge: Cambridge University Press, 2015.

Augustine. *The City of God*. Translated by Henry Bettenson, with a new introduction by G. R. Evans. London: Penguin, 2003.

———. *The Greatness of the Soul, The Teacher*. Translated and annotated by Joseph M. Colleran. New York: Newman, 1950.

———. *The Literal Meaning of Genesis*. https://www.scribd.com/document/159591958/St-Augustine-on-Genesis-1#

———. *On Christian Doctrine*. II.18, e.g. Ch. 18. https://www.ccel.org/ccel/augustine/doctrine.xix_1.html.

———. *The Trinity*. In *Augustine: Later Works*, ed. by John Burnaby. Philadelphia: Westminster, 1955. 17–181.

Bacon, Francis. *The New Organon*. Edited by Lisa Jardine and Michael Silverthorne. Cambridge: Cambridge University Press, 2000.

Baker, Lynne Rudder. *Saving Belief: A Critique of Physicalism*. Princeton, NJ: Princeton University Press, 2017.

Barratt, Geoffrey J. "The Body and You." *The Christian Science Journal*. (January 1995). https://journal.christianscience.com/shared/view/mqtt6r3dw8.

Beauregard, Mario and Denyse O'Leary. *The Spiritual Brain*. New York: HarperOne, 2008.

Beckwith, Francis J. *Politics for Christians: Statecraft as Soulcraft*. Downers Grove, Il: InterVarsity, 2010.

Bedau, Mark. "Cartesian Interactionism." In *Midwest Studies in Philosophy X: Studies in the Philosophy of Mind*, edited by Peter A. French, Theodore E. Uehling Jr., and Howard K. Wettstein, 483–502. Minneapolis: University of Minnesota Press, 1986.

Beilby, James, ed. *Naturalism Defeated? Essays on Plantinga's Evolutionary Argument Against Naturalism*. Ithaca, NY: Cornell University Press, 2002.

Bennett, M. R. and P. M. S. Hacker. *Philosophical Foundations of Neuroscience*. Hoboken, NJ: Blackwell, 2003.

Bloom, Paul. *Descartes Baby: How the Science of Child Development Explains What Makes Us Human*. New York: Basic Books; idem. 2004.

———. "Religion Is Natural." *Developmental Science* 10:1 (2007) 147–51.

Brooks, Paul. "Out of Mind." *Prospect, 109* (April 2005): 1.

Brown, Francis, S. R. Driver, and C. A. Briggs, *A Hebrew and English Lexicon of the Old Testament*. Oxford: Clarendon, 1972.

Brown, Warren S., et al. *Whatever Happened to the Soul? Scientific and Theological Portraits of Human Nature*. Minneapolis, MN: Fortress, 1998.

Churchland, Paul M. *Matter and Consciousness*, revised edition. MIT Press, 1992.

Churchland, Paul M. and Clifford A. Hooker. *Images of Science: Essays on Realism and Empiricism*. Chicago: University of Chicago Press, 1985.

Cohen, S. Marc, and C. D. C. Reeve. "Aristotle's Metaphysics." *Stanford Encyclopedia of Philosophy* (October 2023). https://plato.stanford.edu/entries/aristotle-metaphysics/.

Connell, Richard J. *Substance and Modern Science*. Notre Dame, IN: University of Notre Dame Press, 1988.

Cooper, John W. *Body, Soul, and Life Everlasting: Biblical Anthropology and the Monism-Dualism Debate*. Grand Rapids, MI: Eerdmans, 2000.

Copleston, Frederick. *A History of Philosophy*. New York: Image, 1985.

Corcoran, Kevin. "The Constitution View of Persons." In *In Search of the Soul: Four Views of the Mind-body Problem,* edited by Joel B. Green et al., 153–76. Downers Grove, IL: InterVarsity, 2005.

Crick, Francis. *The Astonishing Hypothesis: The Scientific Search for the Soul*. New York: Simon & Schuster, 1994.

Crouch, Andy. *Culture Making: Recovering Our Creative Calling*. Downers Grove, IL: InterVarsity, 2013.

Dallas Willard Ministries. "Dallas Willard Memorial Service, JP Moreland," YouTube video, 10:14, May 25, 2017. https://www.youtube.com/watch?v=AzSEeIUoksU&ab_channel=DallasWillardMinistries.
Davis, Stephen. "Is Personal Identity Retained in the Resurrection?" *Modern Theology* 2 (July 1986) 329–40.
Dennett, Daniel. *Consciousness Explained*. Boston, MA: Little, Brown and Co, 1991.
Descartes, René. "Meditations on First Philosophy." In *Selected Philosophical Writings*, vol. 2 Translated by John Cottingham et al. Cambridge: Cambridge University Press, 1988.
———. "The Passions of the Soul." In *The Philosophical Writings of Descartes*, vol. 1, translated by John Cottingham, Robert Stoothoff, and Dugald Murdoch, 343–44. Cambridge: Cambridge University Press, 1985.
———. "Sixth Meditation: Of the Existence of Material Things, and of the Real Distinction between the Soul and the Body of Man." In *Discourse on Method and the Meditations*. Translated with an introduction by F. E. Sutcliffe. London: Penguin, 1968.
———. "Treatise on Man." In *The Philosophical Writings of Descartes*, vol. 1, translated by John Cottingham, Robert Stoothoff, and Dugald Murdoch. 107. Cambridge: Cambridge University Press, 1985.
Des Chene, Dennis. *Life's Form: Late Aristotelian Conceptions of the Soul*. Ithaca, NY: Cornell University Press, 2000.
DeWeese, Garrett J. "One Person, Two Natures: Two Metaphysical Models of the Incarnation." In *Jesus in Trinitarian Perspective*, edited by Fred Sanders and Klaus Issler, 144–53. Nashville: Broadman & Holman Academic, 2007.
DeWeese, Garrett J. and J. P. Moreland. *Philosophy Made Slightly Less Difficult: A Beginner's Guide to Life's Big Questions*. Downers Grove, IL: IVP Academic, 2005.
Duvall, Nancy S. "From Soul to Self and Back Again." *Journal of Psychology and Theology* 26 (Spring 1998) 6–15.
Eccles, J. C. *Facing Reality: Philosophical Adventures by a Brain Scientist*. New York: Springer, 1974.
Eccles, J. C., and Daniel N. Robinson. *The Wonder of Being Human*. New York: Springer, 1984.
Elwell, Walter A., ed. "Antinomianism." In *Evangelical Dictionary of Theology*, 57–59. Grand Rapids, MI: Baker, 1984.
———. "Traducianism." In *Evangelical Dictionary of Theology*, 1106. Grand Rapids, MI: Baker, 1984.
Erickson, Millard. *Christian Theology*, 2nd ed. Grand Rapids, MI: Baker, 1998.
Evans, C. Stephen. "Separable Souls: Dualism, Selfhood, and the Possibility of Life after Death." *Christian Scholar's Review* 34 (2005) 327–40.
Foster, John. *The Immaterial Self*. London: Routledge, 1991.
———. "In Defense of Dualism." In *The Case for Dualism*, edited by John R. Smythies and John Beloff, 1–25. Charlottesville: University Press of Virginia, 1989.
Foster, Richard. *Celebration of Discipline: The Path to Spiritual Growth*. London: Hodder & Stoughton, 2008.
France, R. T. *The Gospel According to Matthew: An Introduction and Commentary*. Grand Rapids, MI: Eerdmans, 1985.
Geniusas, Saulius. *The Phenomenology of Pain*. Athens, OH: Ohio University Press, 2020.

Goetz, Stewart. "Substance Dualism." In *In Search of the Soul: Four Views of the Mind-Body Problem,* edited by Joel B. Green et al., 33–60. Downers Grove, IL: InterVarsity, 2005.

———. "A Substance Dualist Response." In *In Search of the Soul: Four Views of the Mind-body Problem,* edited by Joel B. Green et al., 139–42. Downers Grove, IL: InterVarsity, 2005.

———. "Is N. T. Wright Right about Substance Dualism?" *Philosophia Christi* 14, no. 1 (2012) 183–91.

Goetz, Stewart and Charles Taliaferro. *A Brief History of the Soul.* Hoboken, NJ: Wiley-Blackwell, 2011.

Green, Joel B., *Body, Soul, and Human Life: The Nature of Humanity in the Bible.* Grand Rapids: Baker Academic, 2008.

———. *Salvation (Understanding Biblical Themes).* St. Louis, MO: Chalice, 2003.

Green, Joel B., et al., eds. *In Search of the Soul: Four Views of the Mind-Body Problem.* Downers Grove, IL: InterVarsity, 2005.

Groothuis, Douglas. *Christian Apologetics: A Comprehensive Case for Biblical Faith.* Downers Grove, IL: IVP Academic, 2011.

Gross, Charles G. "Geneology of the 'Grandmother Cell.'" *The Neuroscientist* 8 (2002) 512–18.

Grossman, Reinhardt. *The Existence of the World: An Introduction to Ontology.* New York: Routledge, 1992.

Grudem, Wayne A. *Systematic Theology: An Introduction to Biblical Doctrine.* Grand Rapids, MI: Zondervan Academic, 1995.

Guinness, Os. *The Gravedigger File: Papers on the Subversion of the Modern Church.* Downers Grove, IL: InterVarsity, 1983.

Gundry, Stanley N., et al. *Five Views on Sanctification.* Grand Rapids, MI: Zondervan Academic, 1987.

Habermas, Gary and J. P. Moreland. *Beyond Death: Exploring the Evidence for Immortality.* Eugene, OR: Wipf & Stock, 2004.

Habl, Jan. *On Being Human(e): Comenius' Pedagogical Humanization as an Anthropological Problem.* Eugene, OR: Pickwick, 2017.

Hagin, Kenneth E. *Must Christians Suffer?* Tulsa, OK: Faith Library, 1982.

Haldane, J. B. S. "When I Am Dead." In *Possible Worlds and Other Essays.* London: Chatto & Windus, 1927.

Hanson, K. C., and Douglas E. Oakman, translators. "The Theodotus Inscription." http://www.kchanson.com/ANCDOCS/greek/theodotus.htm.

Harré, Rom. *The Philosophies of Science: An Introductory Survey.* Oxford: Oxford University Press, 1972.

Hasker, William. "The Case for Emergent Dualism." In *The Blackwell Companion to Substance Dualism,* edited by Jonathan J. Loose et al., 62–73. Hoboken, NJ: Wiley-Blackwell, 2018.

———. *The Emergent Self.* Ithaca, NY: Cornell University Press, 1999.

———. *Metaphysics: Constructing a World View.* Downers Grove, IL: IVP Academic, 1982.

———. "On Behalf of Emergent Dualism." In *In Search of the Soul: Four Views of the Mind-body Problem,* edited by Joel B. Green et al., 75–100. Downers Grove, IL: InterVarsity, 2005.

———. "Persons and the Unity of Consciousness." In *The Waning of Materialism*, edited by Robert C. Koons and George Bealer, 175–90. Oxford: Oxford University Press, 2010.

Helm, Paul. *Human Nature from Calvin to Edwards*. Grand Rapids, MI: Reformation Heritage Books, 2018.

Hodge, A. A. *Outlines of Theology*. London: Nelson, 1896.

Hodge, Charles. *Systematic Theology, Volume 2*. Grand Rapids, MI: Eerdmans, 1975.

Hoffman, Joshua and Gary S. Rosenkrantz. *Substance: Its Nature and Existence*. London: Routledge, 1997.

Hopp, Walter. *Phenomenology: A Contemporary Introduction*. Routledge Contemporary Introductions to Philosophy. Oxfordshire: Routledge, 2002.

Hughes, Philip Edgcumbe. *The True Image: The Origin and Destiny of Man in Christ*. Grand Rapids, MI: Eerdmans, 1989.

Hume, David. *A Treatise of Human Nature*. Lanham, MD: Prometheus, 1992.

Humphrey, Nicholas. *Soul Dust: The Magic of Consciousness*. Princeton: Princeton University Press, 2012.

Husserl, Edmund. *The Crisis of the European Sciences and Transcendental Phenomenology*. Translated by David Carr. Evanston, IL: Northwestern University Press, 1970.

Imes, Carmen Joy. *Being God's Image: Why Creation Still Matters*. Downers Grove, IL: InterVarsity, 2023.

Ito, Shigehiko, et al. "Performance Monitoring by the Anterior Cingulate Cortex During Saccade Countermanding." *Science* vol. 302, no. 5642 (2003) 120–22.

Johns Hopkins Bloomberg School of Public Health. "Talk Therapy—Not Medication—Best for Social Anxiety Disorder, Large Study Finds." https://publichealth.jhu.edu/2014/talk-therapy-not-medication-best-for-social-anxiety-disorder-large-study-finds.

Johnson, Jan, et al. *Dallas Willard's Study Guide to* The Divine Conspiracy. New York: HarperCollins, 2001.

Kim, Jaegwon. *Philosophy of Mind*, 3rd ed. Boulder, CO: Westview, 2011.

Klassen, Norman, and Jens Zimmermann. *The Passionate Intellect: Incarnational Humanism and the Future of University Education*. Grand Rapids, MI: Baker Academic, 2006.

Koons, Rob. "Against Emergent Individualism?" In *The Blackwell Companion to Substance Dualism*, edited by Jonathan J. Loose et al., 377–93. Hoboken, NJ: Wiley-Blackwell, 2018.

Krapiec, Mieczylaw Albert. *I, Man: An Outline of Philosophical Anthropology*. Translated by Marie Lescoe et al. Boston: Mariel, 1983.

Kripke, Saul. *Naming and Necessity*. Cambridge, MA: Harvard University Press, 1980.

Kuhn, Thomas S. *The Essential Tension: Selected Studies in Scientific Tradition and Change*. Chicago: University of Chicago Press, 1977.

———. *The Structure of Scientific Revolutions*. Chicago: University of Chicago Press, 2012.

Latourette, Kenneth Scott. *A History of Christianity: Beginnings to 1500*, vol. 1. Peabody, MA: Prince, 2007.

Lee, Patrick, and Robert P. George. *Body-Self Dualism in Contemporary Ethics and Politics*. Cambridge: Cambridge University Press, 2008.

Legrenzi, Paolo, and Carlo Umilta. *Neuromania: On the Limits of Brain Science*. Translated by Frances Anderson. Oxford: Oxford University Press, 2011.

Lewis, C. S. *The Collected Letters of C.S. Lewis: Narnia, Cambridge, and Joy, 1950–963*, vol. 3. Edited by Walter Hooper. San Francisco: HarperOne, 2007.

———. "De Descriptione Temporum" Inaugural Lecture from The Chair of Medieval and Renaissance Literature at Cambridge University, 1954. https://files.romanroadsstatic.com/old-western-culture-extras/DeDescriptioneTemporum-CS-Lewis.pdf.

———. *God in the Dock: Essays on Theology and Ethics*. Edited by Walter Hooper. Grand Rapids, MI: Eerdmans, 1970.

———. *Mere Christianity*. New York: Touchstone Books, 1996.

———. *Miracles*. Reprint, Glasgow: Fontana Books, 1963.

———. "On Reading Old Books." https://bradleyggreen.com/attachments/article/97/Lewis.On-Reading-Old-Books.-CS-Lewis.pdf.

———. *The Problem of Pain*. London and Glasgow: Collins Clear-Type, 1940.

———. *Surprised by Joy: The Shape of My Early Life*. San Diego, CA: Harcourt, Brace, Jovanovich, 1966.

———. *Till We Have Faces: A Myth Retold*. New York: HarperOne, 2017.

Loose, Jonathan J. "Materialism Most Miserable: The Prospect for Dualist and Physicalist Accounts of Resurrection." In *The Blackwell Companion to Substance Dualism*, edited by Jonathan J. Loose et al., 470–87. Hoboken, NJ: Wiley-Blackwell, 2018.

Loose, Jonathan J., et al., eds. *The Blackwell Companion to Substance Dualism*. Hoboken, NJ: Wiley-Blackwell, 2018.

Losee, John. *A Historical Introduction to the Philosophy of Science*, 2nd ed. Oxford: Oxford University Press, 1980.

Loux, Michael J. *Substance and Attribute*. Dordrecht: D. Reidel, 1978.

Lowe, E. J. *The Four-Category Ontology: A Metaphysical Foundation for Natural Science*. Oxford: Oxford University Press, 2006.

Luther, Martin. *The Table Talk of Martin Luther*. Edited by Thomas S. Kepler. Mineola, NY: Dover Publications, 2005.

Maslow, Abraham H. "A Theory of Human Motivation." *Psychological Review* 50 (1943) 370–96. https://psychclassics.yorku.ca/Maslow/motivation.htm.

Mathetes. *Letter to Diognetus*. https://www.logoslibrary.org/mathetes/diognetus/06.html

May, Gerald. *Addiction and Grace: Love and Spirituality in the Healing of Addictions*. San Francisco: HarperOne, 2007.

May, William F. *The Physician's Covenant: Images of the Healer in Medical Ethics*. Louisville, KY: Westminster John Knox, 2000.

McDonald, H. D. *The Christian View of Man*. Wheaton, IL: Crossway, 1981.

McGilchrist, Iain. *The Master and His Emissary*. New Haven, CT: Yale University Press, 2009.

McNeill, John T. *A History of the Cure of Souls*. New York: Harper & Brothers, 1951.

Menuge, Angus J. L. "Why Reject Christian Physicalism?" In *The Blackwell Companion to Substance Dualism*, edited by Jonathan J. Loose et al., 395–400. Hoboken, NJ: Wiley-Blackwell, 2018.

Merricks, Trenton. "The Word Made Flesh: Dualism, Physicalism, and the Incarnation." In *The Blackwell Companion to Substance Dualism*, edited by Jonathan J. Loose et al., 452–68. Hoboken, NJ: Wiley-Blackwell, 2018.

Michael, Alexandra Andrea. "Dualism at Work: The Social Circulation of Embodiment Theories in Use." *Signs and Society* S1 (2015) S41–S69.

Miller, Michael Matheson. Interview with Dr. Michael Egnor, *The Moral Imagination*. Podcast audio, October 21, 2020. https://www.themoralimagination.com/episodes/michael-egnor.

Moreland, J. P. "A Christian Perspective on the Impact of Modern Science on Philosophy of Mind." *Perspectives on Science and Christian Faith* 55 (2003) 2–12.

———. *Finding Quiet*. Grand Rapids, MI: Zondervan, 2019.

———. "In Defense of Thomistic-Like Dualism." In *The Blackwell Companion to Substance Dualism*, edited by Jonathan J. Loose et al., 102–22. Hoboken, NJ: Wiley-Blackwell, 2018.

———. "Issues and Options in Exemplification." *American Philosophical Quarterly* 330 (April 1996) 133–47.

———. *The Recalcitrant Imago Dei: Human Persons and the Failure of Naturalism*. London: SCM, 2009.

———. "Restoring the Substance to the Soul of Psychology." *Journal of Psychology and Theology* 26 (Spring 1998) 29–43.

———. "Scientific Late Medieval Aristotelianism (Organicism)." In *The Blackwell Companion to Substance Dualism*, edited by Jonathan J. Loose et al., 105–8. Hoboken, NJ: Wiley-Blackwell, 2018.

———. *Scientism and Secularism: Learning to Respond to a Dangerous Ideology*. Wheaton, IL: Crossway, 2018.

———. "Substance Dualism and the Unity of Consciousness." In *The Blackwell Companion to Substance Dualism*, edited by Jonathan J. Loose et al., 184–207. Hoboken, NJ: Wiley-Blackwell, 2018.

———. "Theories of Individuation: A Reconsideration of Bare Particulars." *Pacific Philosophical Quarterly* 79 (1998) 251–63.

———. "Tweaking Dallas Willard's Ontology of the Human Person." *Journal of Spiritual Formation and Soul Care* 8, no. 2 (2015) 187–202.

———. *Universals*. Oxfordshire: Routledge, 2014.

———. *Universals, Qualities, and Quality-Instances: A Defense of Realism*. Lanham, MD: University Press of America, 1985.

Moreland, J. P., and David Ciocchi, eds. *Christian Perspectives on Being Human: A Multidisciplinary Approach to Integration*. Grand Rapids, MI: Baker, 2015.

Moreland, J. P., and William Lane Craig. *Philosophical Foundations for a Christian Worldview*, 1st ed. Downers Grove, IL: IVP Academic, 2003.

Moreland, J. P., and Scott B. Rae. *Body and Soul: Human Nature the Crisis in Ethics*, 1st ed. Downers Grove: IVP Academic, 2000.

Moreland, J. P., and Stan Wallace. "Aquinas versus Locke and Descartes on the Human Person and End-of-Life Ethics." *International Philosophical Journal* 35, no. 3 (1995) 319–30.

Morris, Leon. *The First Epistle of Paul to the Corinthians: An Introduction and Commentary*. Grand Rapids, MI: Eerdmans, 1985.

Morris, Tom. *If Aristotle Ran General Motors: The New Soul of Business*. New York: Holt Paperbacks, 1997.

Murdoc, Paul. "The Common Denominator of Cultures." In *The Science of Man in the World in Crisis*, 123–42. Ralph Linton, ed. New York: Columbia University Press. 2014.

Murphy, Nancey. "Human Nature: Historical, Scientific and Religious Issues." In *Whatever Happened to the Soul? Scientific and Theological Portraits of Human Nature*, edited by Warren S. Brown et al., 1–30. Minneapolis: Fortress, 1998.

———. "Nonreductive Physicalism." In *In Search of the Soul: Four Views of the Mind-Body Problem*, edited by Joel B. Green et al., 115–38. Grand Rapids, MI: Baker, 2008.

Naselli, Andrew David. *Let Go and Let God? A Survey and Analysis of Keswick Theology.* Bellingham, WA: Lexham, 2010.

Niebuhr, H. Richard. *Christ and Culture.* San Francisco: Harper & Row, 1975.

Noonan, Peggy. "We're More Than Political Animals." *Wall Street Journal* (March 3–4, 2012) A5.

Palmer, Stuart L. "Christian Life and Theories of Human Nature." In *In Search of the Soul: Four Views of the Mind-Body Problem*, edited by Joel B. Green et al., 189–215. Grand Rapids, MI: Baker, 2008.

Papineau, David. *Philosophical Naturalism.* Oxford: Blackwell, 1993.

Pasnau, Robert. *Thomas Aquinas on Human Nature.* Cambridge: Cambridge University Press, 2002.

Pellegrino, Edmund D. and David C. Thomasma. *Helping and Healing: Religious Commitment in Health Care.* Washington, DC: Georgetown University Press, 1997.

Penfield, Wilder. *Mystery of the Mind: A Critical Study of Consciousness and the Human Brain.* Princeton, NJ: Princeton University Press, 1975.

Pinker, Steven. *How the Mind Works.* New York: Continuum, 2006.

Plantinga, Alvin. "Advice to Christian Philosophers." *Faith and Philosophy: Journal of the Society of Christian Philosophers* 1 (1984) 253–71.

———. "Methodological Naturalism." *Perspectives on Science and the Christian Faith* 49 (September 1997) 143–54.

———. *Warrant and Proper Function.* Oxford: Oxford University Press, 1993.

———. *Warranted Christian Belief.* Oxford: Oxford University Press, 2000.

———. *Where the Conflict Really Lies: Science, Religion, and Naturalism.* Oxford: Oxford University Press, 2011.

Plato. "Phaedo." In *The Collected Dialogues of Plato: Including the Letters*, 40–98. Edited by Edith Hamilton and Huntington Cairns. Translated by Hugh Tredennick. Princeton, NJ. Princeton University Press, 2005.

Popper, Karl and John C. Eccles. *The Self and Its Brain.* New York: Springer, 1977.

Rae, Scott B. *Moral Choices: An Introduction to Ethics*, 4th ed. Grand Rapids, MI: Zondervan, 2018.

Reid, Thomas. *The Works of Thomas Reid*, vols. 1–2, 7th ed. Edited by William Hamilton. Edinburgh: Maclachlan & Stewart, 1994.

Renovaré. www.renovare.org.

Rickabaugh, Brandon L. "Responding to N. T. Wright's Rejection of the Soul." *Heythrop Journal* 59 (2018) 201–20.

Rickabaugh, Brandon, and J. P. Moreland. *The Substance of Consciousness: A Comprehensive Defense of Contemporary Substance Dualism.* Hoboken, NJ: Wiley-Blackwell, 2023.

Ross, James. "The Fate of the Analysts: Aristotle's Revenge." *Proceedings of American Catholic Philosophical Association.* 64 (1990) 51–74.

Ruben, Julie A. *The Making of the Modern University.* Chicago: University of Chicago Press, 1996.

Ryken, Leland. *Redeeming the Time: A Christian Approach to Work and Leisure.* Grand Rapids, MI: Baker, 1995.

Ryle, Gilbert. *The Concept of Mind*. New York: Barnes and Noble, 1949.
Satel, Sally and Scott O. Lilienfield. *Brainwashed: The Seductive Appeal of Mindless Neuroscience*. New York: Basic, 2013.
Schutt, Michael P. *Redeeming Law: Christian Calling and the Legal Profession*. Downers Grove, IL: IVP Academic, 2007.
Schwartz, Jeffrey, and Sharon Begley. *The Mind and the Brain: Neuroplasticity and the Power of Mental Force*. New York: HarperCollins, 2002.
Searle, John R. *The Mystery of Consciousness*. New York: New York Review, 1997.
———. "The Problem of Consciousness." *Social Research: An International Quarterly* 60 (1993) 3–16. https://www.sciencedirect.com/science/article/abs/pii/S105381008-3710263?via%3Dihub.
Sellers, Wilfred. *Science, Perception and Reality*. New York: Humanities, 1963.
Shulman, Robert. *Brain Imaging: What it Can (and Cannot) Tell Us About Consciousness*. Oxford: Oxford University Press, 2013.
Siegel, Daniel J. *Mindsight: The New Science of Personal Transformation*. New York: Bantam, 2010.
Simpson, William M. R., et al., eds. *Neo-Aristotelian Perspectives on Contemporary Science*. Oxfordshire: Routledge, 2017.
Singer, Peter. *Practical Ethics*, 3rd ed. Cambridge: Cambridge University Press, 2011.
Sire, James. *The Universe Next Door: A Basic Worldview Catalog*. Downers Grove, IL: IVP Academic, 2020.
Society of Clinical Psychology. "Case Study Mike (Social Anxiety)." https://div12.org/case_study/mike-social-anxiety/#
Spears, Paul D. and Steven R. Loomis. *Education for Human Flourishing: A Christian Perspective*. Downers Grove, IL: InterVarsity, 2009.
Sprigge, T. L. S. *The Importance of Subjectivity*. Oxford: Clarendon, 2011.
Stott, John R. W. *The Letters of John*, Grand Rapids, MI: William B. Eerdmans, 1988.
Swinburne, Richard. "Cartesian Substance Dualism." In *The Blackwell Companion to Substance Dualism*. edited by Jonathan J. Loose et al., 133–51. Hoboken, NJ: Wiley-Blackwell, 2018.
———. *The Evolution of the Soul*. Oxford: Clarendon, 1997.
Taliaferro, Charles. *Consciousness and the Mind of God*. Cambridge: Cambridge University Press, 1994.
———. "Substance Dualism: A Defense." In *The Blackwell Companion to Substance Dualism*, edited by Jonathan J. Loose et al., 49–53. Hoboken, NJ: Wiley-Blackwell, 2018.
Taylor, Jacob Ross. "In His Image and into His Likeness: Human Nature's Theosis in C. S. Lewis's *Till We Have Faces: A Myth Retold*." Honors thesis, Harding University, 2021. https://scholarworks.harding.edu/cgi/viewcontent.cgi?article=1001&context=honors-theses.
Tertullian. *The Soul's Testimony*. Hackensack, NJ: Lighthouse, 2018.
Thacker, Jason. *The Age of AI: Artificial Intelligence and the Future of Humanity*. Grand Rapids, MI: Zondervan Thrive, 2020.
Thompson, Curt. *Anatomy of the Soul: Surprising Connections Between Neuroscience and Spiritual Practices That Can Transform Your Life and Relationships*. Carol Stream, IL: Tyndale House, 2010.
Tooley, Michael. "In Defense of Abortion and Infanticide." In *The Abortion Controversy: A Reader*, edited by Louis P. Pojman and Francis J. Beckwith, 209–33. Boston: Jones & Bartlett, 1994.

Trueman, Carl R. *The Rise and Triumph of the Modern Self: Cultural Amnesia, Expressive Individualism, and the Road to Sexual Revolution.* Wheaton, IL: Crossway, 2020.

———. *Strange New World: How Thinkers and Activists Redefined Identity and Sparked the Sexual Revolution.* Wheaton, IL: Crossway, 2022.

Van Dyke, Christiana. "Not Properly a Person, The Rational Soul and 'Thomistic Substance Dualism.'" *Faith and Philosophy* 26, no. 2 (2009) 186–204.

Van Fraassen, Bas C. *The Scientific Image.* Oxford: Clarendon, 1980.

Van Horn, Luke. "Dualism Offers the Best Account of the Incarnation." In *The Blackwell Companion to Substance Dualism,* edited by Jonathan J. Loose et al., 440–51. Hoboken, NJ: Wiley-Blackwell, 2018.

Van Inwagen, Peter. "I Look for the Resurrection of the Dead and the Life of the World to Come." In *The Blackwell Companion to Substance Dualism,* edited by Jonathan J. Loose et al., 488–500. Hoboken, NJ: Wiley-Blackwell, 2018.

Wallace, Stan W. *Aiding the Christian Scholar in Integrating Faith and Scholarship: How Understanding Constituent Realism Provides Motivation to Respond to Naturalism.* Ann Arbor, MI: ProQuest, 2014.

———. "In Defense of Biological Essentialism." *Philosophia Christi* 10, no. 1 (2002) 29–44.

———. "What Is the Soul, and Why Should We Care? (Part 2)" Discussion with J. P. Moreland, *Thinking Christianly.* Podcast audio, November 15, 2021. https://thinkingchristianly.org/7-what-is-the-soul-and-why-should-we-care/.

Wiggins, David. *Sameness and Substance.* Cambridge, MA: Harvard University Press, 1980.

Wilder, Jim. *Renovated: God, Dallas Willard and the Church That Transforms.* Colorado Springs, CO: NavPress, 2020.

Willard, Dallas. *The Allure of Gentleness: Defending the Faith in the Manner of Jesus.* San Francisco: HarperOne, 2015.

———. *The Divine Conspiracy: Rediscovering Our Hidden Life in God.* San Francisco: HarperOne, 1998.

———. *The Great Omission: Reclaiming Jesus's Essential Teachings on Discipleship.* New York: HarperOne, 2006.

———. "How Concepts Relate the Mind to Its Objects: The 'God's Eye View' Vindicated." *Philosophia Christi,* Series 2, 1, no. 2 (1999) 5–20.

———. *Logic and the Objectivity of Knowledge.* Athens, OH: Ohio University Press, 1984.

———. *Renovation of the Heart: Putting on the Character of Christ.* Colorado Springs, CO: NavPress, 2002.

———. *The Spirit of the Disciplines: Understanding How God Changes Lives.* San Francisco: HarperOne, 1999.

Witherington III, Ben. *Conflict and Community in Corinth: A Socio-Rhetorical Commentary on 1 and 2 Corinthians.* Grand Rapids, MI: Eerdmans, 1995.

Wong, Kenman L., and Scott B. Rae. *Business for the Common Good: A Christian Vision for the Marketplace.* Downers Grove, Il: InterVarsity, 2011.

Woznicki, Andrew. *A Christian Humanism: Karol Wojtyla's Existential Personalism.* New Britain, CT: Mariel, 1980.

Wright, N. T. "Mind, Spirit, Soul and Body: All for One and One for All—Reflections of Paul's Anthropology in his Complex Contexts." Paper presented at the Society of Christian Philosophers Regional Meeting, Fordham University, March 2011. https://ntwrightpage.com/2016/07/12/mind-spirit-soul-and-body/.

———. *The Resurrection of the Son of God*. Minneapolis: Fortress, 2003.
———. *Surprised by Hope: Rethinking Heaven, the Resurrection, and the Mission of the Church*. New York: Harper One, 2008.
Yandell, Keith. "A Defense of Dualism." *Faith and Philosophy* 12 (1995) 551–53.

Selected Subject Index

Abortion, 5, 165–66
Academic discipline, second-order, 46
Amygdala, 16
Anthropology, dichotomist/
　　trichotomist, 88–89
Aquinas, Thomas, 40, 76, 97–99, 102–
　　10, 141–43, 165
Aristotle, 68–69, 80–82, 95, 102–3,
　　109–10, 137, 142–43, 170–72
Attachment love, 29, 51, 65, 71, 112,
　　121–29, 155–56
Athanasian Creed, 26
Atonement, 4, 84, 162
Attributes of God, Communicable, 33
Augustine, Saint, 10, 40, 86, 103, 145
Axons, 17

Binding problem, 67–69, 89
Brain, left and right hemispheres,
　　14–18, 57–58, 65, 105–6, 130,
　　155, 174
Brain, slow and fast tracks, 19, 29, 114,
　　130, 154–56
Brain, triune model (reptilian, limbic,
　　and cortical portions), 65

Capacity (or disposition), 13, 51, 80–
　　108, 124–26, 138, 145–66
Causes, Four (Material, Efficient,
　　Formal, Final), 61, 107, 137–40,
　　173

Cerebrum, 17–19
Cerebral cortex, 17
Cingulate cortex, 16–19, 57, 64, 119
Complementarity thesis, 22–24
Constant conjunction (or correlation)
　　of neural and mental states, 17,
　　43–67, 77, 105, 115
Corpus callosum, 105–6
Council of Constantinople, 26, 40, 88

Dandy-Walker Syndrome, 105
Descartes, Rene, 99, 139, 146
Desires, 20, 39–41, 58, 67, 84, 90–91,
　　101–5, 114, 124, 130, 145, 150,
　　156–62
DNA, 98, 102
Dorsolateral Prefrontal Cortex (PFC),
　　16–19, 30, 58, 65, 67, 74, 176
Downward causation, 60
Dualism, property, 44, 55
Dualism, substance, 45, 107–8, 146

Enlightenment, 20, 28, 69, 116, 138, 145
Epilepsy, 78, 106, 126
Epiphenomena (emergent properties),
　　18, 55–60, 70–77, 140
Epistemology, 53, 115–17, 143, 173
Equality, 5, 167–68
Expressive individualism, 169

Faculties of the soul, 13, 81–105, 148, 156–58
Final resurrection, 11, 28, 31, 34, 75, 94, 142

Gnosticism, 6, 35, 121, 139, 148, 161

Hemispherectomy, 106
Hierarchies of capacities, 83, 148, 152–53
Hippocampus, 16, 67, 74
Hylomorphism, 109–10, 144

Identity, Law of (Indiscernibility of Identicals), 48
Identity theses (Type-Type and Token-Token), 44
Image of God (*imago Dei*), 33, 85–88, 145, 168, 175
Impetus, 114, 155
Incarnation, 4, 162
Incorrigibility, 49
Indexicals, 73, 77
Individuation, 12, 92–93, 145, 166
Infanticide, 166
Intermediate state, 37–39, 50, 75, 104, 108, 142, 151

Keswick view of sanctification, 155
Knowledge argument, 49

Leibniz, Gottfried, 48
Lewis, C. S., 7, 63, 97, 141–43, 175
Libertarian free will, 51
Life Model, 119
Logical fallacy, begging the question, 54, 116, 136
Logical fallacy, false dichotomy, 36, 143–44
Logical fallacy, mereological, 50
Logical fallacy, self-defeating, 45, 70, 77, 117
Logical fallacy, straw man, 144

Materialism: see Physicalism
Mereologically simple, 82, 95
Metaphysically complex, 82

Nephesh, 32, 145
Neurons, 16–17, 30, 44–45, 52, 56–61, 130, 157
Neshama, 32
Neuroplasticity, 15–16, 51–52, 76, 104–5, 127
Neurofeedback, 124
Nicene Creed, 26

Ockham's razor, 133–34
Ontology, 22, 89, 120
Ontological dependency or identity, 56
Ontologically prior, 72

Person, functional definition, 165–66, 175
Phenomenology, 50
Philosophy, Greek, 26, 40, 69
Philosophy, negative role, 81
Philosophy, of science, 45, 134, 137–38
Philosophy, positive role, 82
Physicalism, constitutional, 144
Physicalism, nonreductive, 55–61, 70–77
Physicalism, reductive, 12, 18, 20–28, 40, 44, 53, 57–59, 86, 107, 115, 136, 144–46, 163, 166–68, 173–78
Plato/Platonic/Platonism, 82, 132, 139–42, 148–50, 167–69
Pneuma, 32, 37–38
Prefrontal cortex (PFC), 16–19, 30, 58, 65–67, 74, 176
Principle of Parsimony: see Ockham's Razor
Private access (also first-person perspective), 48–50, 73, 171
Problem of Interaction(ism), 134–44
Proto-Gnosticism: see Gnosticism
Properties, 33, 48–50, 55–62, 71, 82–83, 87, 93–94, 98, 124, 153, 162
Properties, essential, 4, 83–84, 87–95, 138
Properties, degreed, 87
Propositions, 84, 102
Psyche, 32, 145

Quantum mechanics and physics, 53, 135

Relata/Relations, 71–72
Revelation, General, 9–10, 25, 81, 107
Revelation, Special, 10, 25, 81
Ruach, 32

Salvation, 4, 32, 123, 153, 161–63
Sanctification, 86, 158
Scientism, 45–46, 114–18, 137, 143
Sensation, sensory, 46, 62, 68, 89–90, 97–99, 102, 106
Social Anxiety Disorder (SAD), 106–7, 171
Spiritual disciplines (or exercises), 102, 114, 123, 151–59
Substance, primary and secondary, 95
Substance Dualism, Conditional Unity, 110
Substance Dualism, Emergent, 140
Substance Dualism, Holistic, 107–11
Substance Dualism, Integrative, 110
Substance Dualism, Lublin Thomism, 110
Substance Dualism, Platonic/Cartesian, 108, 139–42

Substance Dualism, Thomistic/Thomistic-like/Thomistic hylomorphism, 109–10
Substantial soul, 68, 72, 75, 93–94, 102–7, 110, 126
Synapses, 17, 51, 76
Synecdoche, 32–33

Talk therapy, 104
Teleology (*telos*), 13, 94, 137, 158
Thalamus, 130
Third-person perspective, 49, 73
Total Depravity, 86
Traducianism, 98
Trinity, 26, 82
Triple-bottom line, 172

Unity, functional, 11, 34–36, 39, 97, 103–10, 122, 126, 140
Unity, of consciousness, 66–72, 89
Unity, ontological, 35, 36
Unity, through time, 72–77
University, 28

VIM model of spiritual formation, 114, 154–55

Worldviews, 8, 21, 144, 174

www.ingramcontent.com/pod-product-compliance
Lightning Source LLC
Chambersburg PA
CBHW070252230426
43664CB00014B/2508